Make your reality

THE SPACE PUBLICATIONS GUIDE TO

SPACE CAREERS

Scott Sacknoff & Leonard David

With a Foreword By
BUZZ ALDRIN

Space Publications LLC
Maryland 1998

Guide to Space Careers

By Scott Sacknoff and Leonard David

Published by:
Space Publications
P.O. Box 5752
Bethesda, MD 20824-5752
tel: (301) 718-9603
fax: (301) 718-1837
www.spacebusiness.com

First Printing 1998, Second Printing 1999
Printed in the United States of America
Book Design: Bruce Bennett
Photo Credits: Lockheed Martin, NASA

ISBN 1-887022-05-8

Library of Congress Card Number: 98-060396

10 9 8 7 6 5 4 3 2

Table of Contents

About the Authors

Scott Sacknoff has worked in the industry for more than a decade and is an alumnus of Rensselaer Polytechnic Institute (RPI) and of the International Space University. As an engineer, he worked on a variety of programs and in a number of capacities, including making improvements to the Space Shuttle propulsion system (assembly and test), solving a composite material design problem (laboratory analysis), and improving an enclosed life support system (design). Later, as a business consultant and analyst, he worked with organizations such as NASA and the Department of Defense on their commercialization efforts and with private companies on their strategic marketing plans. In addition, Mr. Sacknoff has authored or co-authored a number of publications, including the annual *State of the Space Industry* and the *United States Space Directory,* as well as performing a number of private studies related to the commercial space industry and its prospects.

Leonard David is a well known industry journalist who has written about the space industry for more than 25 years. He is a frequent contributor to *Space News* and *Aerospace America* and is a former editor of *Final Frontier* magazine. With a personal interest in space exploration and science, Mr. David has become a well known journalist in this genre and maintains numerous contacts within the science, academic, and government communities. As a consultant, he has worked on a number of educational outreach programs developing materials that focus on space-related science projects.

Acknowledgments

This book represents two years of discussions between Scott Sacknoff and Leonard David on determining the best way to help high school and college students learn about the industry while providing them with the resources to find employment and start their careers.

We would like to offer special thanks to the following people for their help with this publication:

- ▲ Barbara Sprungman for acting as a sounding board and providing input on content
- ▲ Andrew Lawler, *Science Magazine*, who provided commentary and editing
- ▲ Tracey Staley, Lockheed Martin, for reviewing this book from the corporate perspective and providing her insights
- ▲ Elizabeth Mezzacappa for seeing the need for this book and sending us her views on "why space?"
- ▲ Bruce Bennett for continued excellence in graphics
- ▲ Buzz Aldrin for providing an insightful foreword
- ▲ Pete Deutermann, a best-selling author, who looked at an early draft and offered suggestions on how to turn a great resource into a good book. His political thrillers are highly recommended. (deuterbook@aol.com)

and last but never least,

- ▲ Mary Stolack, who made endless telephone calls to update addresses and telephone numbers prior to publication.

THE FUTURE AHEAD

by Buzz Aldrin

Embarking on a career in space is a challenging endeavor, one that can surely become an adventure of a lifetime. For me, personally, my experience as an astronaut drew upon skills that I had honed throughout high school, during undergraduate work at West Point,at Air Force fighter pilot school, and then at the Massachusetts Institute of Technology (MIT), where I earned my Doctorate in Astronautics.

That educational foundation launched me into a career at NASA, first as an astronaut in the Gemini program, followed by stepping onto the Moon as a member of the Apollo 11 lunar landing crew. Be assured, there are many more footsteps to be taken. There are more destinations to be reached through robotic and human exploration, given a vibrant and healthy space program. More importantly, you can make a difference in shaping and sustaining the space program during the tomorrows yet to come.

Reflecting back, I can see how things have dramatically changed. In the late 1950s, when Ed White, a friend and later a fellow astronaut, and I started studying astronautics, that term was hardly known. At the time, just a few years had passed after the former Soviet Union lofted Sputnik 1, the first artificial satellite of Earth.

Since those early years, much has taken place—the tremendous growth in satellite telecommunications; the development of Earth-monitoring spacecraft and military reconnaissance and navigation satellites; probes sent outward to Jupiter and Saturn; and the pioneering robotic missions of Viking and, more recently, the Mars Pathfinder and its microrover, Sojourner. Again, there is much progress to be heralded, but even more to consider in future years.

My timing was just right for stepping into the space program. I asked myself what should I do with my experience as a fighter pilot and expertise in guidance and navigation. Back then, joining objects in space was something that was only done by computer analysis or discussed in academic circles through a paper or two. In fact, I could count on the fingers of one hand experts in the field of space rendezvous. I took a hard look at rendezvous methods and boiled them down to simple terms, in essence, a seat-of-the-pants approach.

The merit of my rendezvous ideas gradually gathered adherence. It afforded the pilots—the astronauts—a way to have back-up charts, make measurements, and assure that we were contributing to a spacecraft's flight. The essence of my rendezvous techniques was tested during Gemini, then applied to the Apollo lunar landing effort, and is still used today. I am very proud of that contribution.

For those of you just starting on a high-technology career path, I can offer some observations that may prove helpful. They have been valuable to me.

I came up with a philosophy several years ago of trying to go through life with my arms outstretched to cut a wide swath. Gather in as many things as possible into your realm of awareness. You will be surprised at what you encounter and collect.

Don't focus too narrowly. Open up. Change direction. Look at a problem from as many perspectives as enter into your mind. Add to somebody's idea or bring two people together who are working the same problem, but from different points of view. Polish your ideas as much as possible, then try them out on others. Don't be afraid to give away a little bit in order to become a cooperative person. To say you need an open mind is obvious.

Regarding work skills, the fact that you are reading these words already indicates your ability to research and stay current. You should keep abreast of space projects underway in the private sector, in academic settings, and in the government. Maintaining your knowledge of world space progress is very important. By reading publications such as *Space News*, *Via Satellite*, or the *State of the Space Industry* and joining space advocacy groups such as the National Space Society and the Planetary Society or technical groups such as the American Institute of Aeronautics & Astronautics, ASPRS, and others, you can maintain a sense of what issues are most relevant to your work.

By the way, there are two other skills that are very critical. Written and verbal communication skills are a must. I've seen many a creative idea, prompted by brilliant deduction, stymied because he or she could not convey to others their thought in a coherent or persuasive manner. Furthermore, project into the future and consider how your creative viewpoint might contribute to the betterment of conditions today. Think about how best to take advantage of the microgravity, the vacuum, the three dimensional freedom, the view from a distance that space offers?

Lastly, the most powerful tool you can acquire or experience is an ability to give away what you have just learned, to teach someone the basics of what you are attempting to understand.

There is much to do in assuring that humankind moves forward in its space pursuits. For me, I am sharply focusing my energies on reusable first-stage rocketry, the opening up of a space tourism market, and the creation of a sustainable space program. I envision future use of recycling cycling orbits that form a network supporting inner-solar system traffic of cargo, men and women that ply routinely between the Earth and Mars. Beautiful simplicity coupled with a ballet of intricate celestial mechanics are the hallmarks of this approach. Reusable, recyclable space transportation is the key to our future. Fashioning a sustainable space program is paramount to becoming a true, spacefaring civilization.

I encourage you to read through this book. You will note that a diversity of potential space careers is available to you, as are the types of industries involved. Best wishes for selecting the discipline that suits your interests, and good luck in contributing to the space program in the years ahead.

It will take more than rhetoric to assure that a strong and imaginative 21st century space program becomes reality. We must remain steadfast in our resolve to create new economic opportunity in space, assure the integrity and security of our home planet Earth, return to the Moon, establish a foothold on Mars, and move humanity outward into the Universe at large.

You can become a major contributor to that future.

Buzz Aldrin
Apollo 11 Astronaut
Chairman of the Board, the National Space Society
April 1998

1 INTRODUCTION

"One small step for man, one giant leap for mankind"[1]
"In a galaxy, far, far, away..."[2]
"Space, the final frontier..."[3]

These words have inspired billions of people around the Earth, and the great space race to the Moon in the 1960s enabled many to live out their dream. Today, hundreds of thousands of people from all walks of life and all backgrounds have chosen to work in a space industry that is far larger and more diverse than ever before.

By choosing a career in the space industry, you can be part of today's vision—whether it means bringing the world closer together through communications systems, monitoring and preserving the Earth's climate and resources, or building rockets and satellites to explore the solar system, the galaxy, and the wider universe.

Though many do not realize it, the space industry is much bigger and broader than human space programs such as the Space Shuttle, Mir, and International Space Station. In fact, according to recent statistics, the space industry generates more than 15 times the revenues of Hollywood films. Other statistics show that the commercial space industry has been one of the fastest growing sectors of the economy, with parts of it surpassing 20% growth annually. With this growth comes jobs. A recent estimate forecast that more than 40,000 new jobs are being created each year—one of which could be filled by you.

PROSPECTS FOR FINDING A JOB—
HOW IS THE INDUSTRY DOING?

Recent statistics show that the space industry surpassed $97 billion in revenue in 1998 and is expected to exceed $130 billion by the year 2002.

Most of this money comes from two sources. The first is government-sponsored research, development, and operational programs, including NASA programs such as the Space Shuttle, the International Space Station, and Hubble Space Telescope, and those related to procuring and operating military spacecraft for communications and intelligence gathering. The second source is the commercial telecommunications satellite industry, which enables everything from international telephone calls to the distribution of cable and television programming to local providers.

While government budgets likely will remain stable for the foreseeable future, commercial satellites focusing on telecommunication applications are expected to be the major industry driver for the next ten years. New satellite applications enabling direct-to-consumer sales have already started; with the first of these products—direct-to-home television—becoming the fastest selling consumer electronics product in history.

Other consumer products and services derived from space systems expected to begin operation in the next ten years include precision farming, enhanced search and rescue operations, vehicle theft prevention, tracking of non-violent criminals, and portable telephones that work as well from a small mountain village in remote areas of the Himalayas as they do in downtown Los Angeles.

With $50 billion of new growth anticipated in the next five years, a dramatic need for employees is forecast. Already many firms, especially satellite manufacturers and component suppliers and companies involved with software, information processing, and data analysis, say that they are having difficulty finding qualified or trainable employees—and they are expecting that this shortfall could get worse. This does not guarantee you will find a job in the space industry, but it is always better to look in a growing and dynamic field.

WHAT WILL THIS BOOK DO FOR ME?

By understanding an industry, you will be more aware of the opportunities within it. This book is meant to provide you with an overview of the space business and act as a resource guide to assist you in your employment search. Each chapter focuses on a different area of knowledge:

- ▲ Chapter 2: What is the Space Industry?
- ▲ Chapter 3: History
- ▲ Chapter 4: Where are We Today? Where are the Opportunities?
- ▲ Chapter 5: How is the Industry Organized? What is the Role of Industry; of Government; of the Military?
- ▲ Chapter 6: What Does the Space Industry Do? What are the Different Types of Positions/Careers Available?
- ▲ Chapter 7: What Universities Specialize in Space? How Can I Increase My Employment Prospects? What Are Some Scholarships and Fellowships That Are Available?
- ▲ Chapter 8: Where Can I Find Information on Open Positions? Should I Consider Contract or Temporary Employment? What Resources are Available in Starting My Own Company?
- ▲ Chapter 9: Why, How and Where Should I Network?
- ▲ Chapter 10: On Becoming an Astronaut
- ▲ Appendix A: Addresses of Companies
- ▲ Appendix B: Key Space Terms and Acronyms

WHAT IS THE SPACE INDUSTRY?

The space industry is as diverse as any that exists in today's economy. Its participants are involved in everything from engineering and science to manufacturing, farming, geology, and meteorology.

The space industry can actually be thought of as a combination of other industries—communications, electronics, information technology, computers and software, environmental monitoring, manufacturing, medical and biotechnology research, etc. In fact, many of the people who are actually involved in or who are using the industry in their work, consider themselves part of other industries. Ask a medical researcher what they do and chances are they will say that they are in the medical field, not someone involved with evaluating pure crystals manufactured in the microgravity of space.

An overview of the activities that encompass the space industry is presented in Chapter 2. Take notice that many of the activities of the space industry actually take place on the ground.

WHAT TYPE OF ENGINEERING OR SCIENCE BACKGROUND SHOULD I PURSUE?

The space industry has technical positions available for people with all types of degrees and backgrounds. Some of these are:

Electrical Engineering
Materials Engineering
Information Technology
Meteorology
Biology
Optics & Electro-Optics
Chemical Engineering
Astrophysics and Astronomy
Fluid Mechanics
Mechanical Engineering
Computer Sciences
Physics
Software Design
Physiology
Telecommunications
Chemistry
Celestial Mechanics
Aerospace Engineering

According to an executive at Lockheed Martin, recent college graduates with degrees in computer science and information technology, electrical engineering, and physics are highly sought after.

ARCHITECTS, ARCHAEOLOGISTS, GEOLOGISTS

Professionals in a number of other fields use products derived from the space industry. Architects and civil planners use satellite data and digital maps to locate, evaluate, and understand sites. Geologists use satellite data to find and interpret possible mineral deposits. Archaeologists have recently begun using space data to locate sites located under hundreds of feet of sand or on the ocean bottom. In the not-too-distant future, farmers may draw on satellite data to learn about the health of their crops, but also which areas need fertilizer or watering. They may also use Global Positioning System satellites to automate their tractors and harvesters.

I DON'T HAVE A TECHNICAL DEGREE. CAN I FIND SOMETHING WITH MY BACKGROUND?

Many of the people who work in the space industry do not have technical backgrounds. Careers exist for people with skills in:

Business	Sales and Marketing	Accounting
Technical Writing	Journalism	Graphic Design
Video Production	Conference Management	Public Relations
Administration	Employee Relations	Economics

WHAT IF I DON'T LIVE NEAR A NASA FACILITY?

Although there may be a higher concentration of companies involved with space in some parts of the United States, companies and organizations involved with space and telecommunications can be found throughout North America and the world. A review of the organizations listed in Appendix A shows a diversity of firms located throughout the United States.

DREAMING THE FUTURE— WHAT ARE THE OPPORTUNITIES OF TOMORROW?

Much of the commercial expansion in the space industry has occurred within the last decade. Within the next 20 years, space assets will revolutionize communications and video and data services and provide us with a new understanding of our planet and our environment. Within the next 50 years, many within the industry believe that we will:

- ▲ Land on Mars
- ▲ Perform research on orbiting space platforms with revolutionary results
- ▲ Establish a permanent base on the moon
- ▲ Harness the sun's energy with solar power stations
- ▲ Expand flight opportunities to the average citizen

The key to these latter visions is the dramatic reduction in the cost to launch materials and people into space. Should this occur, the industry could grow in directions not dreamed of today.

You can be a part of that growth.

1 Neil Armstrong's first words upon setting foot on the Moon
2 Quote from Star Wars © and ™ Lucasfilm, Ltd.
3 Star Trek and related elements © and ™ Paramount Pictures Corporation

2 WHAT IS THE SPACE INDUSTRY?

"If I could get one message to you it would be this: The future of this country and the welfare of the free world depends upon our success in space."
—President Lyndon Baines Johnson

WHAT OR WHERE IS SPACE?

There is no agreed-to legal definition of where space begins. Different groups have their own definitions about where the Earth's atmosphere ends and space begins. Scientists have classified space as beginning approximately 200-300 nautical miles above the Earth's surface. Aeronautical engineers generally use 54 nautical miles because aerodynamic forces are negligible above this height. The U.S. Air Force defines space as heights above 44 nautical miles.

In conventional and customary law, the major space powers generally accept "the lowest altitude attained by orbiting space vehicles" as the threshold of space. This point, the lowest at which a satellite can maintain a stable orbit, is 90 nautical miles. To maintain this altitude, a satellite must have a minimum velocity of 17,500 miles per hour parallel to the surface of the earth.

Ninety miles up, the temperature is as low as -450F or 3K; gravity is less than 1/1000 of that which would be felt on Earth; and radiation is much higher and more damaging than that on Earth.

But this harsh environment offers a unique perspective on the Earth below. Like looking off of the top of a tall building or a mountain and gazing across the land below, the vantage point of space offers an opportunity to scan the Earth's surface, thereby providing a platform for remote sensing, weather monitoring, intelligence gathering, and communications.

THE SPACE INDUSTRY IS...

Telecommunications

(Using spacecraft to relay data from one part of the Earth to another)

Cable Programming Distribution
Live Television and Video Transmission
International Telephony
Mobile and Wireless Communications
Messaging Services
Telemedicine
Tele-Education
High-Speed Internet Access
VSAT Private Communication Networks
Direct-to-Consumer Video and Radio

Spacecraft Manufacturing

(Construction of the satellites)

Telecommunications Satellites
Remote Sensing Satellites
Planetary Exploration Satellites
Weather Satellites

Launch Vehicles

(Rockets used to place payloads in orbit)

Expendable Launch Vehicles
Reusable Launch Vehicles
Space Shuttle Operations

Ground Equipment

(The equipment on Earth that is used to receive and/or transmit data to and from spacecraft)

Ground Stations
Computer Software and Hardware
Electronic Receiving and Transmission Equipment
Antennas
High Capacity Data Storage
Information Technology

Ground Operations

(Facility design, development, and use. Monitoring and controlling spacecraft or launch vehicles)

Satellite Operations
Telemetry and Control Hardware and Software
Health Monitoring and Operations Planning Software
Component Test Facilities
Launch Vehicle Spaceports

Global Positioning System Services

(The use of a 24-satellite system developed by the U.S. military that provides accurate positioning data anywhere on Earth)

Enhanced Air Traffic Control
Directional Services for Automobiles
Improved Search and Rescue Devices for Boaters & Hikers

Remote Sensing

(The monitoring of the Earth using space-based sensors)

Weather Prediction and Forecasting
Monitoring of the Earth's Environment
Searching for Natural Resources
Analysis of Soil and Land Conditions for Farming
Use of Digital Terrain Maps
National Security Intelligence Gathering

Human Space Activities

(Activities related to the human exploration of space and the effects of long and short-duration spaceflight on the human condition)

Space Station
Space Shuttle
Medical, Physiological, and Psychological Research

Microgravity

(Use of the special environmental conditions in space—such as low temperature and low gravity—to develop materials or products)

Production of New or Improved Materials
Enhanced Crystals for Biomedical Research
Biomedical Drug Development

Space Science

(The study of the universe including stars, planets, interstellar materials, as well as the effect of the space environment on the Earth)

Astrophysics and Astronomy
Astrodynamics
Cosmology
Astrobiology

Technology Research and Development

(A number of technologies are used in various aspects of the industry)

Optics
Power Systems
Propulsion Systems
Lasers
High-Temperature Materials
Composite Materials
Thermal Control
Guidance, Navigation, and Control
Robotics

Future Space Activities

(Many of these have been proposed or forecast to occur within the next 50 years)

Permanent Lunar Bases
Human Missions to Mars
Manufacturing in Orbital Facilities
Mining of Nearby Asteroids
Orbital Solar Power Generation Stations
Toxic and Nuclear Waste Disposal
Tourism

Support Services

Administrative Support
Technical Support
Legal and Licensing
Financial Services
Media and Publishing
Satellite, Launch Vehicle, and In-Orbit Insurance

Working in the Space Industry

BIOGRAPHY #1

Name: Dino Lorenzini, President and CEO

Organization: SpaceQuest, Ltd.

Responsibilities

As the leader of a small entrepreneurial company, I am responsible for articulating the vision of the company, setting the strategic direction, and bringing together the people, partners, resources, and financing needed to turn the vision into reality.

Background

B.S. Engineering Science (US Air Force Academy)
S.M. Astronautical Engineering (Massachusetts Institute of Tech)
Sc.D. Astronautical Engineering (Massachusetts Institute of Tech)
MBA Business Management (Auburn University)

Career Path

I began my career as an Astronautical Development Engineer in the Air Force with increasing responsibilities for the management of space programs. I finished my 23-year career in the Air Force as the Program Manager for DARPA's space-based laser program and as Director of the Space Systems Architecture Study for the Strategic Defense Initiative Program. In 1989, I made the transition into commercial space to become a part of a revolutionary movement of lower cost systems with unlimited business potential. I have worked primarily with low-Earth orbit communication satellite systems as the chief architect, system designer, business developer and promoter.

Why Space?

I chose space over aviation because I was impressed by the accomplishments of the first astronauts during my college studies, motivated by the challenge to be creative and innovative, and inspired by the national importance and significance of the US space effort at that time.

Words of Wisdom

Try to think outside of the box, don't accept negative responses, be persistent in the pursuit of your dreams, and work hard.

HISTORY

Visionaries and the Government Dominate the pre-1986 Industry

"It is difficult to say what is impossible, for the dream of yesterday is the hope of today is the reality of tomorrow."—Robert Goddard

SCI-FI HISTORY IN THE MAKING

To boldly go where no one has gone before...a saying deeply rooted in the popular movie and television presentations of Star Trek™.

But in truth, that premise of space exploration has been alive and well for centuries, from the earliest beginnings of science fiction writing to the theater screens of today. The use of imagination as propulsion can transport a person outward to the moon or at warp speed to faraway stars.

Take for instance Cyrano de Bergerac, who in the early 1700s wrote a story making use of rocket propulsion to commute to the Moon. Over a hundred years later, writer Edward Everett Hale detailed in the *Atlantic Monthly* what is thought to be the first fictional account of a space station. In 1869 and 1870 issues of the magazine, Hale concocted a tale of a large brick satellite housing 37 adventurers.

Cruising at a cosmic altitude high above Earth, Hale's whimsical postulations had the crew of the brick satellite aiding navigating seaman below. To communicate with sailors, the brick moon's residents jumped up and down on the exterior of the satellite in Morse code fashion: long bounds for dashes, short leaps for dots!

Similar in manner, but based on more solid technical footing, were the writings of Jules Verne. In his classic 1865 novel, *From the Earth to the Moon,* he wrote of a bullet-shaped space ship resembling, in many ways, the launch vehicles of today. Verne painted a picture of space travel featuring conditions commonly encountered by 20th century human explorers. His notion of rocket propulsion, however, was to utilize a huge cannon to fire a passenger-riding projectile to the Moon.

On the heels of Verne, was the H.G. Wells 1897 account that detailed a Martian invasion of Earth—the *Independence Day*™ of its time—*The War of the Worlds*. The fanciful tales of Verne and Wells were soon not so fanciful. The technology of flight, first by aircraft and then by rockets, jumped from fictional verbiage to high-velocity hardware.

It was science fiction that helped seed the imaginings of many a "real" rocketeer.

PIONEERING IMAGINEERS

Like the harnessing of the atom, the erection of the Grand Coulee dam or the construction of the Panama Canal, spaceflight symbolizes technological achievement in the 20th century. Russian scientist and father of astronautics Konstantin Tsiolkovsky; German scientist Hermann Oberth; America's Robert Goddard; and consummate space visionary Wernher von Braun, to name but a few, shaped the foundation of thought on space. Each plotted out a masterplan for utilizing space. Each developed a blueprint for opening up the frontier of space, predicated not on fanciful fiction but on the mathematical and scientific knowledge of the day. These individuals were part of a vanguard of visionaries who turned dream machines into reality.

Called the father of American rocketry, Goddard was first inspired by Jules Verne's *From the Earth to the Moon*, as well as the writings of H.G. Wells. He earned his science degree from Worcester Polytechnic Institute in Massachusetts and, as a young engineer, received his first patent for a "rocket apparatus" in 1914. Shortly thereafter, as a part-time physics professor at Clark University, he crafted his first rockets. It was in this period that Goddard wrote his seminal paper submitted to the Smithsonian Institution and published in 1919: *A Method of Reaching Extreme Altitudes*.

It was Goddard, at a time when Oberth and Tsiolkovsky theorized space futures, who began putting metal into the sky by the late 1920s.

His inventions were modest in the beginning. The world's first liftoff of a liquid-fueled rocket, in fact, took place from the snow-covered Massachusetts farm of Goddard's Aunt Effie. That rocket flight on March 16, 1926, lasted a little more than two seconds, shooting to an altitude of 184 feet. A few years later, seeking a site with good weather for year-round testing and distant from populated areas, Goddard continued his pioneering work in Roswell, New Mexico, where he remained until 1941. From fuel pumps to guidance and control gyroscopes, Goddard pursued many innovations that would later become mainstay technology for all large rockets.

Robert Goddard's impressive accomplishments, along with the growing body of work by Tsiolkovsky and Oberth, inspired many others around the globe. Among them was Wernher von Braun.

At the close of World War II, the U.S. Army seized the top engineers of Germany's rocket effort, including von Braun. He and his engineering team were brought to the U.S. and stationed at the Army's White Sands Proving Grounds in New Mexico, along with some captured V-2 rockets. It was from this talent base that America's space program was formed.

To most Americans in the late 1940s and early 1950s, space travel was relegated to the backs of cereal boxes and the latest installment of Flash Gordon in the Sunday newspaper. But von Braun sought to challenge people's incredulity by detailing a scientifically accurate, step-by-step approach to space exploration in the pages of popular magazines.

Keen on technical accuracy, von Braun stressed that space travel did not require huge leaps in new technology. Wheel-shaped space stations, an expedition to the planet Mars, even the hauling into Earth orbit the necessary fuel and hardware via a "space ferry"were envisioned by von Braun and explained in matter-of-fact detail as eminently possible.

Part politician, part salesman, full-time rocketeer, von Braun began to captivate millions with his space blueprint.

In early October 1957, people found space travel on their front doorstep. Newspaper headlines screamed that the Soviet Union had hurled Sputnik-1, the first artificial satellite, into space. The Space Age had arrived.

BUILDING FROM THE GROUND UP

Sputnik-1's launch on October 4, 1957, proved to be a propaganda coup for the Soviet Union. Even more striking was the orbiting of Sputnik-2 just four months later. It carried the first living organism --a dog named Laika-- into space.

Following the Soviet Union's one-two space punch came America's response. Sitting atop its booster on December 6, 1957, at a Cape Canaveral, Florida launch site, was the tiny Vanguard satellite. It was all over in one second. On live television, the satellite fell to the ground as part of flaming booster wreckage, later to be found still beeping amidst the rubble.

"American Sputnik goes Kaputnik!" complained more than one newspaper headline after the failure. American pride and prestige went up in flames with the satellite.

Vanguard's failure forced President Dwight Eisenhower to order a U.S. Army team to loft a satellite into orbit within 90 days. That team was headed by Wernher von Braun.

The U.S. Explorer 1 satellite rocketed into orbit on January 31, 1958. At just 30 pounds, the U.S. Explorer satellite weighed just 1/36th the weight of the more massive Sputnik 2. In those early, heady days of U.S. and Soviet one-upmanship, bigger meant better, never mind the scientific utility of a satellite.

For many Americans, those first Soviet satellite successes signaled something akin to a technological Pearl Harbor in space. With the lofting of America's Explorer 1, a two-nation "space race" was on.

In early 1958, the U.S. Congress passed the National Aeronautics and Space Act, signed into law on July 29 by President Eisenhower. America's commitment to space was born out of the government-industry partnerships the nation had made for aeronautical research. The National Advisory Committee for Aeronautics (NACA) was transformed into the National Aeronautics and Space Administration (NASA). At the same time, a largely classified military space program began to grow in the Pentagon.

Almost one year to the day after Sputnik-1 began circling Earth, NASA officially started to orchestrate the nation's civilian space agenda. By the end of 1960, NASA had some 19,000 employees in its ranks. As its partnerships with industry and academia grew, so too did plans for a stable of launch vehicles and various types of application satellites and scientific probes.

Experimental Earth remote sensing, weather, navigation, and communications satellites were in orbit by the early 1960s. In the communications arena, NASA's leadership spawned the first operational telecommunications satellite of its type, Telstar 1, built by Bell Laboratories. Primitive compared to the communications satellites now in use, Telstar-1 relayed up to 60 telephone calls or a single television channel simultaneously. A few months after Telstar-1 was launched in July 1962, The Radio Corporation of America (RCA) began operating its Relay 1 communications satellite. Television networks, also in their infancy in many ways, boasted of programs that were "Live Via Satellite".

From the vantage point of space, satellites afforded new ways to monitor crops, watch for dangerous weather conditions, transmit voice, data and images around the globe, as well as maintain a vigil for trouble spots that might jeopardize national interests.

But early in the 1960s, it was the combination of politics and romance of the cosmos that supplied the U.S. space program with its most difficult challenges.

THE HUMAN TOUCH

Standing before Congress on May 25, 1961, President John F. Kennedy set America on a course to the Moon.

"I believe that this Nation should commit itself to achieving the goal, before this decade is out, of landing a man on the Moon and returning him safely to Earth.

No single space project in this period will be more important for the long-range exploration of space; and none will be so difficult or expensive to accomplish..."

That bold decision was made a month after the Soviet Union's Yuri Gagarin had become the first human to orbit the Earth. Moreover, Kennedy's commitment came just 20 days after a U.S. astronaut had flown a 15-minute "suborbital" flight of a Mercury space capsule. That test shot, a quick, albeit highly televised mission, set in motion America's human spaceflight program that remains active today.

Kennedy's declaration about putting a man on the Moon had provided, in a very real way, a finish line for the "space race" between two superpowers.

There was no question that reaching for the Moon's terra incognita would be daunting. Kennedy himself addressed that fact in September 1962, noting before a 35,000 person gathering at Rice University in Texas:

"We shall send to the Moon, more than 240,000 miles from the control center in Houston, a giant rocket more than 300-feet tall, made of new metal alloys, some of which have not yet been invented, capable of standing heat and stresses several times more than have ever been experienced, fitted together with a precision better than the finest watch, carrying all the equipment needed for propulsion, guidance, control, communications, food, and survival, on an untried mission to an unknown celestial body..."

Apollo became a work in progress. To reach for the Moon demanded the harnessing of pilot skills and hardware through the Mercury and Gemini space missions. From single-seater flights of Mercury astronauts to the two-seat Gemini spacecraft, these piloted space missions around the Earth would provide the nation the necessary wherewithal to reach outward a quarter of a million miles.

But moving from rhetoric to real rocketry meant calling upon government, industry, and university skills. The Apollo effort would consist of more than 20,000 companies employing almost 400,000 people throughout the country.

On July 20, 1969, Earth's collective heartbeat sped up, then the world experienced a heartstopping moment. A strange shadow fell across the Moon's cratered, grayish landscape. Gliding above a stark vista, billions of years old, an oddly shaped craft hovered in mid-vacuum, its landing legs outstretched. Apollo 11's lunar lander, the Eagle, settled down in a place called the Sea of Tranquility. The goal set by President Kennedy less than nine years earlier had been met.

U.S. astronauts Neil Armstrong and Buzz Aldrin became the first human visitors to another world, as fellow Apollo 11 astronaut Michael Collins orbited the Moon in an Apollo command module.

As the two moonwalkers explored the lunar surface, history was written in a place Aldrin calls "magnificent desolation".

Starting in July 1969 and continuing until December 1972, six expeditionary crews visited the Moon, allowing 12 men from Earth, and a nation, to reach beyond their grasp.

AFTER APOLLO

While placing humans on a distant world was the vision of generations, the dream was short-lived. The efforts of 400,000 government workers and hundreds of aerospace contractors and suppliers were left in the lunar dust when Apollo 17 astronauts departed the Moon in December 1972.

Financial belt tightening by the government led to the cancellation of Apollo 18, 19, and 20. Within a decade, the space agency's budget fell from more than $20 billion in 1964 to $6 billion. With the decline in NASA's funding, visionary schemes of large space stations, bases on the Moon, and sending human expeditions to Mars faltered. Layoffs swept through the industry as the Apollo lunar landing program ended.

During the budget problems of the 1970s, NASA continued its mission by focusing on robotic planetary explorers and using leftover Apollo space hardware.

Among the many automated missions which NASA pursued during this time were Pioneer, Voyager, and the Viking mission to Mars, each of which proved a major success by relaying to Earth valuable data on our solar system and our planetary neighbors.

NASA also pursued two other major programs in this period. Skylab, America's first space station, was crafted by modifying a huge Saturn V upper-stage left over from the Apollo program, and the Apollo-Soyuz Test Project, which led to the first docking in space between a Russian and U.S. spacecraft.

The Skylab space station was lofted into orbit in May 1973. Once in space, Apollo spacecraft were launched to the Earth-circling complex. From May 1973 into November of that year, three separate Apollo spacecraft, each carrying a three-person crew, were lobbed to the Skylab outpost. Years later, abandoned, it re-entered the Earth's atmosphere.

The Apollo-Soyuz Test Project was also designed from ex-Apollo hardware and it signaled the end of the "space race" between the Soviet Union and the United States. High above Earth, a two-person cosmonaut crew linked their Soyuz spacecraft with the three-seater Apollo on July 17, 1975. The docking in orbit was enabled by specially designed hardware, a forerunner of the equipment now in use to couple the space shuttle with Russia's Mir space station. Handshakes in space and some joint scientific experiments were part of the high-altitude détente that took place over two days. That cooperation lapsed for two decades before renewal.

WINGING A WAY TO ORBIT

From 1975 until 1981, the United States astronaut corps was essentially grounded. Technical snags, budgetary squeezes, and a largely disinterested Congress combined to stretch out the development of the U.S. Space Shuttle.

The $10 billion investment in the Space Shuttle program resulted in an initial fleet of orbiters: Enterprise (used for air drop tests only), Columbia, Discovery, Atlantis, and Challenger.

As billed in the 1970s, the Space Shuttle program would be used to ferry materials to and from a space station; resupply, repair, recover, and deploy satellites; as well as act as a winged laboratory in space. The first flight in 1981 proved successful, but the program proved far costlier than NASA predicted.

Then, 25 flights later, the Challenger orbiter and its seven crew members were lost en route to orbit. On January 28, 1986, just 73 seconds into flight, a leak in the joint of one of the two solid rocket propellant motors led to the destruction of Challenger and the astronauts. Following more than two years of remaking the shuttle program, Discovery winged its way into space on September 29, 1988. This mission paved the way for the dozens of shuttle launches that have since followed.

One outcome from the tragic Challenger disaster was the rebirth of a private expendable rocket fleet. With the space shuttle launching commercial and government payloads, the production and launch of Delta, Atlas, Titan and other expendable launch vehicles diminished in the early 1980s. With the shuttle fleet grounded for those two years, payloads stacked up on Earth. This helped the commercial launch vehicle industry to establish itself as the primary means for placing payloads into space.

TOWARD A PRIVATE EYE ON SPACE

While the seeds for commercial space endeavors were planted in the 1950s, it is only in the last few years that many of the early applications in communications and Earth monitoring have emerged as economic powerhouses. Industry officials and executives now view space primarily as a place for business and commerce, rather than exploration.

History reminds us that the quest to push back frontiers here on Earth begins with exploration and discovery, followed by settlement and economic development.

Recall the epic journey of Lewis and Clark, sparked by the acquisition of new land through the Louisiana Purchase. Remember the investment made a generation later for what was then termed the "frozen wasteland" called Alaska. Since the earliest times, expansion into new regions is almost always met by skepticism. The risk of a return on investment is high, with patience and fortitude required to win a payoff.

Today, space is paying off. New goods and services for individuals and businesses here on Earth result from space commerce. Tapping the unique vantage point of space has led to a wide array of telecommunications services, to prospecting for new energy sources, and to better management of agricultural resources. The vacuum and microgravity conditions found in Earth orbit may lead to improved manufacturing processes or techniques to produce superior electronic components and life-saving drugs.

The use of space for economic expansion is not only being explored by U.S. interests but also by countries around the world. More than 30 nations are actively pursuing space-related opportunities. Japan, France, Germany, Russia,

China, Canada, Brazil, Australia, to name a few, have recognized the tremendous market potential for commercial space operations. Today, space is no longer just the province of superpowers—it's a key component of the global marketplace.

Understanding history puts you one step closer to understanding the present.

Working in the Space Industry

BIOGRAPHY #2

Name: Elizabeth Silbolboro Mezzacappa, Ph.D.

Organization: NIMH Post-doctoral Fellow, Columbia Presbyterian Medical Center

Responsibilities

Psychological Research

Background

B.A. in Psychology and Biology (University of Pennsylvania)
Ph.D. in Medical Psychology (Uniformed Services University of the Health Sciences)
International Space University Life Sciences Program

Career Path

I am a space medical psychologist working among the space, academic, and lay communities. While I am not in daily contact with space medical and psychological research, I see it as part of my responsibility to inform these communities through my participation in invited workshops, lectures, and presentations on psychological aspects of living in space. For example, I founded the Aerospace Biomedical Association at my university and have presented research relating to space at the Ford Foundation/National Academy of Sciences meeting and the annual convention of the American Psychological Society. My Ph.D. work was supported by a NASA Graduate Student Researchers Fellowship award. Early last year, I was awarded a small grant from the 2111 Foundation for Exploration to investigate cloistered contemplative communities as psychosocial space analogs.

Why Space Psychology?

Human psychological and behavioral issues are perhaps the least examined, yet arguably the most important for a permanent exodus into space.

Words of Wisdom

You don't have to be "in" the space community per se, to make a contribution to the space effort in psychology. One can do meaningful space psychological research outside of NASA or the space industry. There is so much to be done, it can't be restricted to the typical spaceflight centers.

4 WHERE ARE WE TODAY?

The U.S. government still remains the largest single customer and client for space hardware, software, and services. However, by the late 1990s, revenues generated from commercial activities surpassed government expenditures for the first time. Much of this revenue resulted from the use of satellites for telecommunications and video transmission.

According to the latest *State of the Space Industry,* the industry is expected to surpass $120 billion in revenues by the year 2000, almost $50 billion in new growth within the next five years.

Global Space Industry Revenues
(Forecast in Billions U.S. Dollars)

	1998	1999	2000	2001
Infrastructure	55,471	57,806	59,506	61,676
Telecommunications	33,646	37,368	42,214	48,700
Emerging Applications	4,648	5,905	7,518	9,654
Support Services	3,827	3.933	3,686	3,732
Satellite Manufacturing*	13,753	13,425	13,432	13,504
Launch Vehicles*	5,820	7,100	6,825	6,674
Ground Segment*	22,499	23,635	25,501	27,975

*Included in Infrastructure Totals

*From the 1999 State of the Space Industry

THE INDUSTRY'S LARGEST CUSTOMER

Space remains a vital national interest to the U.S. government for both civil and military activities. The military and intelligence communities continue to integrate space assets and information into their everyday activities. To highlight its importance, speeches by Air Force officials have stated that the organization is actually the "Air and Space Force".

The government is also the prime mover in a number of major civilian programs, including the International Space Station, the Earth Observation System, and a host of smaller science-oriented missions, including our first Pluto-bound spacecraft. In addition, research performed with the Hubble Space Telescope and other space observatories is expected to continue to provide amazing data on the universe.

While budgets for civil government, military, and intelligence space activities are expected to remain flat for the foreseeable future, they will still provide a hefty $40 billion annually.

TELECOMMUNICATIONS—COMMUNICATING FINANCIAL SUCCESS

The driving force in the industry today is the use of satellites for telecommunications. Satellites provide an instant infrastructure, enabling companies to provide services quickly and, in most cases, less expensively than developing a fiber or cable network. New and expanded telecommunications services are fueling the growth in satellite manufacturing, ground station hardware and software, and launch vehicles.

In 1997, officials at the Federal Communications Commission stated that they anticipated that the $550 billion telecommunications industry could expand to almost $1 trillion by the year 2000.

Some of this growth is fueled by current and emerging space-based services including:

- ▲ Direct-to-home television and radio broadcasting
- ▲ Fixed satellite services allowing the broadcast and distribution of television and cable programming
- ▲ Mobile telephony services providing users access to communications regardless of where they travel—whether it is over the ocean, in the mountains, or downtown in a major city
- ▲ International voice and data traffic
- ▲ High-speed Internet access
- ▲ Store-and-forward communications which allow a remote or mobile site to transmit or receive data or location information, such as text messages, electric utility usage, or the current location of a trucking fleet
- ▲ Transmission of medical graphical data and information
- ▲ Transmission of educational programming

Among the companies involved with the space-based telecommunications sector are satellite providers, such as PanAmSat, Intelsat, Iridium, Globalstar, and Columbia Communications; satellite manufacturers who operate their own satellites either internally or through joint business arrangements such as Hughes Communications, Loral Space & Communications, and Lockheed Martin; satellite users, such as HBO, Turner Broadcasting, and TCI; and teleport operators providing uplink and downlink services, such as the Washington International Teleport or Teleport Minnesota.

SATELLITE MANUFACTURING—BUILDING THE BRAINS

With a high demand for communications satellites and other spacecraft, companies involved with the assembly and integration of satellites and the manufacture and design of satellite components are extremely busy. Many have indicated that they are having trouble finding enough employees to fill demand. Based on the number of potential orders that have been announced, they expect to have this problem for the foreseeable future. The satellite manufacturing sector is dependent on a number of factors including the ability of their potential customers to secure the financing needed to build, launch, and operate their satellites. Not every venture reaches fruition. In addition, the sector is cyclical—like many other manufacturing industries there is a lag between when the satellite is delivered and when a new satellite is ordered to replace the existing one. For example, a number of forecasts have predicted that there will be a slight downturn in the early part of the next century. However, this should not affect employment within the broader sector as the long-range forecast shows steady growth and a long-term need for trained and qualified workers.

LAUNCH VEHICLES—PROVIDING ACCESS TO SPACE

The expendable launch vehicle, which is based on the missile technologies of the 1950s, is the mainstay of the space industry for placing payloads in orbit. Vehicles such as the Delta, Atlas, and Titan have been launching payloads into space on a regular basis since the 1960s.

In addition to the above, a range of expendable vehicles, which are available on the market today, are capable of launching small, medium, intermediate, or heavy payloads (4000kg+). Among these are a number from international sources. In addition to the above vehicles are the Ariane (Europe), Proton (Russia), Sea Launch (U.S./Ukraine), Cosmos (Ukraine), Long March (China), Taurus (U.S.), Athena (U.S.), and more than 30 other vehicles capable of launching a payload into orbit or a suborbital trajectory.

With the boom in telecommunications and Earth-monitoring satellites, launch vehicles are booked several years in advance. The driving forces for the industry today are reliability and cost reduction. Every launch vehicle manufacturer seeks to improve its performance and reliability and reduce payload losses. With the competitive nature of the industry, these companies undergo a constant evaluation process to reduce costs, improve profit margins, and provide a better price to their customers.

By the late 1990s, several companies began investigating the concept of developing a reusable launch vehicle to reduce payload costs dramatically. The Space Shuttle and the Russian Buran vehicle (which was flown once and put in storage) are the only vehicles today which can be flown, inspected, and then reused. The Space Shuttle never met its price target. New reusable vehicles, proposed by Lockheed Martin (VentureStar/X-33), Kistler Aerospace, Kelly Space & Technology, Pioneer Rocketplane, Rotary

Rocket Corporation, and a number of other companies, are currently under development with initial testing slated to occur around the turn of the century.

While the technical success and actual cost of operating these vehicles still remains to be proven, many believe that reusable vehicles could dramatically affect the launch vehicle industry before the year 2005.

This should not affect your choice in a career or a company. Keep in mind the following: expendable and reusable vehicles will co-exist for a number of years; a number of technologies and skills learned on expendable vehicles are applicable to reusable vehicles; and, perhaps most importantly, many experts forecast that a significant reduction in the cost to launch material into space will dramatically increase the number of payloads and in turn, the workforce needed to meet demand.

GROUND STATIONS, OPERATIONS, AND EQUIPMENT

The ground segment of the industry is comprised of a diverse group of activities from monitoring spacecraft operations to manufacturing electronic equipment to designing buildings and facilities.

More companies are involved in this sector than in manufacturing spacecraft and launch vehicles, and it is a major source for industry revenue and employment. In fact, if you look at the revenues associated with the ground segment (presented earlier in this chapter), you will notice that it is larger than the infrastructure associated with the space segment.

Among the many functions of the ground segment are:

- ▲ Monitoring vital signs of spacecraft in orbit
- ▲ Receiving and transmitting data to and from satellites for communications or remote monitoring of the earth
- ▲ Manufacturing electronic and mechanical equipment for the above activities
- ▲ Testing launch vehicle and satellite hardware
- ▲ Developing and testing software to plan and monitor the satellite

- ▲ Manufacturing portable electronics to receive voice, data, video, or GPS signals from satellites
- ▲ The design and assembly of launch, test, and operations facilities

With a number of commercial applications due to expand rapidly in the next 10 years, notably direct-to-home television and radio, mobile satellite telephones, and a wide range of consumer electronics using GPS receivers, the ground segment is expected to experience continuing growth.

REMOTE SENSING— UNDERSTANDING THE EARTH FROM ABOVE

Next to telecommunications, one application poised to achieve rapid growth in the next decade is the use of remotely sensed data. From their vantage point high above the Earth, space-based sensors scanning the Earth are providing users on the ground with an unprecedented array of information. With customers in the agricultural, mining, and oil exploration industries, to name a few, the potential number of users for this data is enormous.

Multi-spectral sensors onboard Earth-observing spacecraft produce data helpful in monitoring yields from global wheat production, tracking oil pollutants on the oceans, keeping an eye on river flooding, identifying locales of valuable minerals, and watching urban and rural growth patterns. As an example, the annual plotting of snowcover in the western U.S. can cost one-half million dollars to process the data, but it saves almost $50 million annually in hydroelectric and irrigation efficiencies.

Between now and 2005, more than 20 new satellites are planned to be launched into space to focus their high-resolution sensors or radar imaging systems on the Earth, dramatically increasing the amount and type of data that is available and the need for data interpretation.

Revenues from Earth monitoring are expected to surpass $12 billion annually by the year 2010 and to provide a new generation of users with information on improving and better utilizing the resources on our planet.

As colleges and universities from around the world begin to integrate remote sensing analysis and usage courses into their programs in forestry, oceanography, farming, civil planning, and even archaeology, remote sensing data will become integrated into everyday jobs, much like the computer, the fax machine, and e-mail.

GPS—NAVIGATING THE EARTH

Yet another illustration of space commerce is the growth and rapid acceptance of satellite navigation. The first navigation satellites were orbited in the early 1960s by the U.S. Navy to increase the target accuracy of submarine-launched missiles.

Today's Global Positioning Satellite System (GPS) consists of 24 U.S. Department of Defense satellites equally distributed like a net around the Earth. They provide military and civilian users with worldwide position, velocity, and time information data whose accuracy is dependent upon the equipment used on the ground. By the second half of the 1990s, commercial, portable GPS receivers began to become more widely used in aircraft, boats, and automobiles.

A number of firms now offer hand-held GPS navigation units that are inexpensive, highly portable, and easy to operate. This burgeoning commercial space tool is in wide use in oil tankers, aircraft, and even as part of a backpacker's take along items. In coming years, more and more cars are likely to offer built-in GPS units, not only to help a driver efficiently move from point to point, but also to help locate stolen vehicles. Additionally, civil aircraft, under the direction of the FAA, will use GPS as the principal form of navigation—dramatically improving safety.

In the next 10 years, the GPS industry is projected to increase its revenues by several hundred percent, from $3 billion in 1997 to more than $10 billion by early in the next decade. With an estimated 700,000+ users in 1995, the number of devices utilizing GPS is expected to exceed five million by the year 2005.

MICROGRAVITY—MADE IN SPACE

Both commercial as well as government firms are looking at harnessing space for manufacturing "made-in-space" products. The exotic nature of space—specifically its microgravity environment—appears to be of benefit in making semiconductor materials, higher strength alloys, even pharmaceuticals. By reducing the forces of gravity on industrial processes, a diverse menu of products is hoped for. Research already conducted in space is pointing the way to such possibilities—products that are purer or exhibit qualities of greater value over their Earth-made counterparts.

Making the most of microgravity means removing the gravity factor during formation of materials. Experiments conducted in orbit show that manipulating temperature, composition, and fluid flow can be controlled far better, thanks to the microgravity condition of space.

A drawback to such experimentation, however, remains the high cost of access to orbit. Expensive space transportation costs are focusing commercial attention on low-volume, low-weight, high-value products, like pharmaceuticals, electronic components, optical devices, and metal alloys.

For the moment, activities in this area have been limited. Research activities are expected to expand as opportunities for getting "time in space" increase—first with services provided on the Mir Space Station, the European free-flyer Eureka, and the Space Shuttle, and later with the International Space Station and commercial platforms.

As opportunities for research expand, numerous products of commercial value are anticipated. A sample of these include:

- ▲ ELECTRONIC MATERIALS: Pure, nearly perfect crystals are required in computers and numerous optical and electronic devices. Crystals, made of silicon or other material, are the basis for semiconductors, infrared sensors, and lasers. Space processing permits crystal purity and uniformity far beyond those possible on Earth.

- ▲ METALS, GLASSES, AND CERAMICS: High-strength metals and temperature-resistant glasses and ceramics are essential to power generation, propulsion, aviation, aerospace, and related applications. Containerless processing in space permits the mixing and solidification of metals and ceramics in forms and at levels of purity that cannot be attained on Earth.

▲ BIOLOGICAL MATERIALS: Separation of macro-molecules (Proteins, enzymes, cells, and cell components) is fundamental to all fields of biological research. Collagen fibers to replace injured human connective tissues, and urokinase—a drug for countering blood clots—are two biological substances that can be produced in space. Many more candidate pharmaceuticals have been tagged as suitable for manufacturing in microgravity, with greater efficiency, lower cost, and in greater purity levels than can be achieved on Earth.

NEW HORIZONS

During the next several decades, expanded commercial use of space is expected. Growth of telecommunications satellite services, including navigation, disaster warning, fleet dispatch, and emergency location will spawn 21st century firms specializing in customer products. Earth remote sensing products tailored to a customer's interests will become widespread around the globe. So too will navigation, detection, and tracking services.

Eventually, advances in launch systems should bring about more routine and far less costly access to space. If so, more commercial firms are likely to find business niches in Earth orbit. Mining, manufacturing, power generation, and hazardous waste processing, have all been proposed.

A new, next generation of launchers could one day herald airline-like space operations. This would be a major step forward in reducing the cost of placing payloads and people into orbit. Some experts predict a large, blossoming space tourism travel business! Privately operated spaceliners may drop off passengers at "get away from it all" recreational and resort facilities in low-Earth-orbit. This is not a far-fetched vision if the expense of lofting a passenger drops to the same level as taking a luxury ocean cruise.

Given the pioneering spirit that opened up previous frontiers here on Earth, our newest arena for economic expansion may be as unbounded as space itself.

Words of Wisdom

Consider looking at both large and small organizations. Most graduates naturally look at high-profile companies, many of which tend to be mid-sized or larger firms. Seeking out smaller companies or those that don't promote themselves as often may be more time-consuming, but they may provide an opportunity that you couldn't find elsewhere.

Words of Wisdom

Your first job may not reflect the career you end up with. Many people will tell you how their backgrounds and interests evolved over the years, and their careers and responsibilities changed as opportunities arose.

Working in the Space Industry

BIOGRAPHY #3

Name: Christopher Silva, Management Consultant

Organization: A.T. Kearney

Responsibilities

As space consultant of a leading management consulting firm, I conduct industry analysis, market expansion, profit improvement, pricing, and investment services to leading space companies.

Background

M.A. Economics (Tufts University)
MBA Business Management (Averett College)

Career Path

I began my career as an Air Force Officer working with high-technology applications of space assets. After six years, I left the Air Force to become the Business Development Director for GRCI, a high technology company specializing in technical consulting to U.S. Government customers. My activities included support to the Strategic Defense Initiative Program. I received an MBA while at GRCI. In 1993, I made the transition into management consulting and specialized in commercial space. My focus was to help my clients improve their business through reduced cost structures, enhanced/proactive market awareness and improved manufacturing processes, and through the development of business cases to attract investment.

Why Space?

I chose space because I believed it is the future of high-technology business. Commercialization of this almost limitless resource affords companies worldwide the opportunity for growth and the betterment of mankind.

Words of Wisdom

If your job benefits others and you care about what you do, it ceases to be a job.

5 THE STRUCTURE OF THE SPACE INDUSTRY

"No one person, no one company, no one government agency, has a monopoly on the competence, the missions, or the requirements for the space program"
— President Lyndon Baines Johnson

Public companies, private companies, government agencies, military organizations, universities, and research institutions all play a role in the space industry. Understanding the roles and activities of each will enable the job hunter to identify opportunities.

In addition to organizing the space industry by activity, such as satellite manufacturing, launch vehicles, ground stations, telecommunications, remote sensing, etc., segments of the industry are also commonly grouped by funding source, customer, or client. These groupings are:

- ▲ Commercial Space
- ▲ Civil Space
- ▲ Military Space
- ▲ Academic Community

The military space sector in the United States is comprised of activities performed by the Air Force, Navy, Army, and other defense and intelligence organizations, such as the National Reconnaissance Office, the National Imagery and Mapping Agency, the CIA, etc.

The civil space sector in the United States is primarily comprised of activities from NASA, NOAA (National Oceanic and Atmospheric Administration), the Federal Aviation Administration, the Department of

Commerce, and the Federal Communications Commission. Agencies such as the Environmental Protection Agency (EPA), Department of the Interior, and the Federal Emergency Management Agency (FEMA), also maintain official space budgets, mainly for procuring remotely sensed data.

COMMERCIAL SPACE SECTOR

Commercial space has historically been defined as those projects or programs in which the commercial entity has acted as the lead in the financing of a new venture. With many companies focusing on both commercial and government clients, the commercial sector also includes all those private or public corporations that derive their revenues from government contracts but whose employees are private sector workers. Many of the organizations that focus only on government clients are commonly known as government contractors.

The commercial sector performs most of the work within the space industry including almost all of the manufacturing. Within this sector is an entire network of companies -- from manufacturers and suppliers, to distributors and marketers of the product or resource, to those that provide technical and professional support.

For instance, to place a communications satellite in orbit requires companies that assemble the launch vehicle; manufacture the components (fuel pumps, tank structure, etc.) and subcomponents (gaskets, seals, valves, etc.) which make up the launch vehicle; provide test personnel to make sure the vehicle will operate correctly; provide ground launch personnel to monitor the vehicle prior to and during flight; and supply systems integrators that mate the payload to the vehicle. This does not even include positions such as the truck driver who delivers the tanks of fuel to the launch site, any of the positions associated with designing, building, testing, or evaluating the satellite, the components within the satellite, the payload on the satellite, or the support personnel who will continually monitor the performance and data stream from the satellite once it reaches orbit.

Likewise, other organizations that would have an interest in this flight would be risk evaluators who provide the insurance on the vehicle and the satellite; the financiers who funded the project; the market researchers who evaluated the system's potential; and the marketing personnel who sell the data derived from the satellite.

In particular, the commercial sector is being driven by orders from companies providing telecommunications services. Fixed satellite services continue to form the core of the industry providing telephony, video, and data services. Beginning in 1997, mobile communication satellite networks such as Iridium and Globalstar were being launched into orbit, marking a new age in communications. Mobile communications services allowing worldwide access to telephony, data, and paging, even in remote areas, will change the way we communicate.

By the early part of the next century, expanded use of satellites for high-speed access to the Internet, telemedicine, telephony services in remote areas, and distance learning will drive the industry to even greater heights.

WHO IS THE COMMERCIAL SPACE SECTOR?

Many of the best known institutions that make up the space industry historically have been large manufacturers who are better known for their work in aviation or as government contractors—Boeing, Lockheed Martin, Hughes, Loral, and TRW, to name a few.

During the last several years,as the commercial space industry has expanded, these firms and others have begun not only to compete for commercial orders but also have internally financed several commercial ventures. As an example, Lockheed Martin, Hughes, and Loral are financing and operating a number of commercial communications satellites in addition to manufacturing them, and Boeing has privately financed the development of the Sea Launch and Delta III launch systems.

Keep in mind that, like the rest of the U.S. economy, small and mid-size companies play a major role in the industry and represent the major area of job growth.

With the emergence of new commercial applications in the 1980s and the expansion of the commercial sector in the 1990s, as many as 2000 small and mid-sized companies may be involved in the space industry. Many of these are less than 20 years old and are involved with communications services, ground equipment and electronics, or technical and professional support. Companies such as Orbital Sciences Corporation, Iridium, Globalstar, PanAmSat, and Analytical Graphics have grown to become major players in the industry and are, or may become, household names.

WHERE ARE THEY LOCATED?

With the expansion of the telecommunications industry and the growth associated with data distribution for remotely sensed imagery, space-related companies can be found in all 50 states and throughout the world.

While spread out across the U.S., like other industries, space organizations concentrate in areas where the money is. Many of the companies involved with space infrastructure are based in California, Texas, Colorado, Florida, and the Washington D.C. area—areas close to major NASA and military research and/or operational facilities.

In contrast, companies that are involved with the application of space-derived data are scattered throughout the nation with no discernible pattern—sometimes they are near their primary users, other times simply where the president of the firm sets up shop.

CIVIL SPACE SECTOR

The U.S. government is involved in a wide range of non-military activities from monitoring the Earth's weather to human and robotic exploration of the solar system.

The lead organization for civil space activities is NASA. With an annual budget of more than $13 billion, NASA is the lead civilian agency for advancing the state-of-the-art in technology and hardware development, the utilization of space data and resources, and science and space exploration.

Several other federal organizations have an interest in space. For instance, weather satellite operations, forecasting, and research are handled by NOAA, the National Oceanic and Atmospheric Administration, an agency that is part of the Department of Commerce. Additionally, the Departments of Interior and Agriculture as well as FEMA, the Federal Emergency Management Agency, use a tremendous amount of remotely sensed data to achieve their missions. Other agencies are responsible for regulatory issues. The Federal Aviation Administration, for example, oversees the safety of commercial launch facilities and vehicles, while the FCC is responsible for regulating communications.

Federal Agency Space Budgets
Budget Authority in Millions—Real Year Dollars

Agency	Activity	1986 ($ million)	1990 ($ million)	1996 ($ million)
NASA	science, communications, remote sensing, launch vehicles, human space flight	7,807	12,324	13,884
Dept. of Interior	remote sensing	2	31	35
Dept. of Commerce (including NOAA)	remote sensing, weather forecasting, trade promotion	309	243	430
Federal Aviation Administration	launch regulation	0	4	6
Environmental Protection Agency	remote sensing	0	5	6
National Science Foundation	research grants	106	125	147
Dept. of Energy	power systems	35	79	46
Dept. of Defense (Unclassified budget)	communications, remote sensing, launch vehicles, data analysis, ground operations	14,126	15,616	11,514

CIVIL SPACE: NASA

While many think of NASA as a well-funded agency, its budget is actually modest compared to other parts of the government. Its $13+ billion budget is small compared to the $200+ billion dollar budgets of the Department of Defense or the Department of Health and Human Services. In relation to the total U.S. budget, NASA's funding has dropped in recent years to about one percent of total government expenditures.

NASA's major programs for the 1990s and beyond are the Space Shuttle, Space Station, the Hubble Space Telescope, the Mission to Planet Earth, and a host of space science missions.

Major areas of NASA interest

- ▲ Science including astrophysics, medical research, and microgravity
- ▲ Human space flight including the Space Shuttle and Space Station
- ▲ Advanced technology development
- ▲ Transfer of NASA developed technology to the overall economy

NASA ORGANIZATIONAL STRUCTURE

NASA is a single agency of the U.S. government comprised of a headquarters and several field centers. NASA Headquarters is located in Washington D.C. and is responsible for interactions with Congress, the White House, and other government agencies. The headquarters staff exercises management control over the centers and other installations. Responsibilities include overseeing and developing new programs and projects; establishing management policies, procedures, and performance criteria; and evaluating the progress of all phases of the aerospace program.

NASA Headquarters
300 E Street, SW
Washington, DC 20546
(202) 358-0000

NASA FIELD CENTERS

Most of the technical work within NASA is performed and managed at its field centers. Located throughout the United States, each center maintains specializations and facilities in certain technical areas. Interestingly enough, the field centers overlap in their technical expertise and have been known to compete with each other for certain projects and funding.

Field Center Locations

▲ Ames Research Center	California
▲ Goddard Space Flight Center	Maryland
▲ Kennedy Space Center	Florida
▲ Langley Research Center	Virginia
▲ Lewis Research Center	Ohio
▲ Marshall Space Flight Center	Alabama
▲ Stennis Space Flight Center	Mississippi
▲ Johnson Space Center	Texas
▲ Jet Propulsion Laboratory	California
▲ Wallops Flight Facility	Virginia

NASA AMES RESEARCH CENTER

Ames is located in Moffett Field, California near Palo Alto. The center specializes in computer science and applications, computational and experimental aerodynamics, flight simulation, flight research, hypersonic aircraft, rotorcraft and powered-lift technology, aeronautical and space human factors, life sciences, space sciences, solar system exploration, airborne science and applications, and infrared astronomy.

Ames Research Center
Moffett Field, CA 94035-1000
(650) 604-5000

NASA GODDARD SPACE FLIGHT CENTER

Goddard Space Flight Center (GSFC) is located in Greenbelt, Maryland just outside of Washington D.C. The Center is involved in, among other things, research in the Earth and space sciences and the design, fabrication, and testing of scientific satellites that survey the Earth and the universe, as well as tracking satellites and suborbital space vehicles.

Because of its versatility, Goddard scientists can develop and support a mission, and Goddard engineers and technicians can design, build, and integrate the spacecraft. Goddard is also involved in implementing suborbital programs using small and medium expendable launch vehicles, aircraft, balloons, and sounding rockets.

The scientific data from these and other space flight experiments are catalogued and archived at the National Space Science Data Center at Goddard in the form of magnetic tapes, microfilm, and photographic prints to satisfy the thousands of requests each year from the scientific community.

Goddard is also the lead center for the Earth Observing System (EOS) missions, which will address pressing global issues, such as the depletion of atmospheric ozone and long-term global warming.

Goddard Space Flight Center
Greenbelt Road
Greenbelt, MD 20771-0001
(301) 286-2000

JET PROPULSION LABORATORY

JPL, operated by the California Institute of Technology in Pasadena, is a government-owned contractor-operated research, development, and flight center that performs a variety of tasks for NASA. The laboratory is engaged in exploring the Earth and the solar system with robotic spacecraft. In addition to the Pasadena site, JPL manages the Deep Space Communications Complex, a station of the worldwide Deep Space Network located at Goldstone, Calif. The Deep Space Network allows for spacecraft communications, data acquisition and mission control; for the study of space with radio science; and for performing basic and applied scientific and engineering research in support of the nation's interests.

Among the NASA flight projects under JPL management are the Voyager, Galileo, Magellan, Ulysses, and Topex/Poseidon spacecraft and the Mars Pathfinder mission. Major space science instruments include the wide field/planetary camera for the Hubble Space Telescope, the NASA scatterometer, and the Shuttle imaging radar.

Jet Propulsion Laboratory
4800 Oak Grove Drive
Pasadena, CA 21109-8099
(818) 354-4321

NASA JOHNSON SPACE CENTER

The Johnson Space Center, located in Houston, Texas, is the focal point for NASA's human space flight program. It is responsible for the management and operation of the Space Shuttle and the Space Station and develops and maintains excellence in the fields of project management, space systems engineering, medical and life sciences, lunar and planetary geosciences, and crew and mission operations. JSC is responsible for developing and integrating scientific, medical, and technological experiments and payloads, which are flown on the Space Shuttle, Spacelab, the International Space Station, and the Russian Mir Space Station.

Lyndon B. Johnson Space Center
NASA Road One
Houston, TX 77058-3696
(281) 483-0123

NASA KENNEDY SPACE CENTER

Kennedy Space Center (KSC), located on the east coast of Florida, was established in the early 1960s as the launch site for the Apollo lunar landing missions. The Center is NASA's primary facility for the test, checkout, and launch of payloads and space vehicles. This includes launch of manned vehicles at KSC and oversight of NASA missions launched on unmanned vehicles from Cape Canaveral Air Force Station, Florida and Vandenberg Air Force Base in California. The center is responsible for the assembly, checkout, and launch of the Space Shuttle vehicles and their payloads, landing operations, and turnaround of Shuttle orbiters between missions, as well as preparation and launch of vehicles from Vandenberg.

John F. Kennedy Space Center
Kennedy Space Center, FL 32899-0001
(407) 867-7110

NASA LANGLEY RESEARCH CENTER

Located in Hampton, Virginia, Langley's primary mission is to perform basic research in aeronautics and space technology. Major research fields include aerodynamics, materials, structures, flight controls, information systems, acoustics, aeroelasticity, atmospheric sciences and nondestructive evaluation. Approximately 60 percent of Langley's work is in aeronautics.

For space, Langley researchers performed extensive work on the structure, aerodynamics, and thermal protection system for the Space Shuttle. Langley also manages an extensive program in atmospheric sciences, seeking a more detailed understanding of the origins, chemistry, and transport mechanisms that govern the Earth's atmosphere.

Langley Research Center
Hampton, VA 23665-0001
(757) 864-1000

NASA LEWIS RESEARCH CENTER

Located in Cleveland, Ohio, Lewis is NASA's lead center for research, technology, and development in aircraft propulsion, space propulsion, space power, and satellite communications. Lewis is also active in microgravity research and has several facilities, including drop towers, among its resources.

Lewis Research Center
21000 Brookpark Road
Cleveland, OH 44135-3191
(216) 433-4000

NASA MARSHALL SPACE FLIGHT CENTER

Located in Huntsville, Alabama, Marshall Space Flight Center, under the direction of Wernher von Braun, designed and developed the launch vehicles that carried man to the Moon in the Apollo program.

Today, launch vehicle development is but one aspect of the center's multi-faceted operation, which includes expertise in materials processing in a microgravity environment, materials testing, and manufacturing.

Marshall had a significant role in the development of the Space Shuttle and continues to manage the Space Shuttle main engines, the external tanks that carry liquid oxygen and liquid hydrogen for those engines, and the solid rocket boosters that, together with the engines, lift the Shuttle into orbit.

The center also has a key role in the development of scientific payloads and experiments to be flown aboard the Shuttle. The center operates NASA's Spacelab Mission Operations Control Center and a Payload Crew Training Complex for those missions.

The center also managed the development and initial orbital checkout of the Hubble Space Telescope, the Advanced X-ray Astrophysics Facility, and the Inertial Upper Stage and Transfer Orbit Stage vehicle systems.

George C. Marshall Space Flight Center
Huntsville, AL 35812-0001
(205) 544-2121

NASA STENNIS SPACE CENTER

NASA's John C. Stennis Space Center (SSC), located in Stennis, Mississippi, has grown over the past 30 years into NASA's premier center for testing large rocket propulsion systems for the Space Shuttle and future generations of launch vehicles. Since 1975, SSC's primary mission has been the research and development and the flight acceptance testing of the Space Shuttle main engines.

Stennis personnel also are involved in scientific research, remote sensing technology and applications, and technology transfer. The center has been designated as NASA's lead center for the commercialization of remote sensing technology and, as such, works with the public and private sectors to expand the use of remote sensing imagery and technology.

John C. Stennis Space Center
Stennis, MS 39529-6000
(601) 688-2211

NASA WALLOPS FLIGHT FACILITY

Part of the Goddard Space Flight Center, the Wallops facility is located on Delmarva Peninsula off the coast of Virginia. Wallops manages and implements NASA's sounding rocket program, which uses solid-fueled rocket launch vehicles to accomplish approximately 35 scientific suborbital missions each year. Wallops also manages and coordinates NASA's Scientific Balloon Program using thin-film, helium-filled balloons. Wallops plans and conducts Earth and ocean physics, ocean biological and atmospheric science field experiments, satellite correlative measurements, and developmental projects for new remote sensor systems. The main thrust of this effort is in support of the Laboratory for Hydrospheric Processes.

Wallops Flight Facility
Goddard Space Flight Center
Wallops Island, VA 23337-5099
(757) 824-1000

OTHER CIVIL SPACE ORGANIZATIONS

NATIONAL OCEANIC AND ATMOSPHERIC ADMINISTRATION (NOAA)

NOAA's activities relate to the monitoring, prediction, research, and distribution of data related to the weather and the environment. Scientists at NOAA are active in research related to meteorology, oceanography, solid-earth geophysics, and solar terrestrial sciences. In addition, NOAA maintains and employs personnel for ground station operations of satellites and archiving and distributing large databases containing research and meteorological data. NOAA's facilities can be found throughout the United States and include the National Hurricane Center in Florida, the National Climactic Data Center in North Carolina, the Space Environment Center in Colorado, and the Satellite Operations Control Center in Maryland.

NOAA
Silver Hill and Suitland Road
Suitland, MD 20746
(301) 457-5113

DEPARTMENT OF COMMERCE

The International Trade Administration—Office of Aerospace, works to promote and expand opportunities for U.S. companies on the international market.

International Trade Administration
Room 2128
Washington, DC 20230
(202) 482-1229

FEDERAL AVIATION ADMINISTRATION

The Office of Commercial Space Transportation is responsible for regulating and promoting commercial space launch activities and commercial launch facilities. The Licensing and Safety Division reviews launch vehicles and launch facilities for public safety and environmental impact. The Space Policy Division develops and analyzes policies and market trends to help with guidelines for free and fair trading practices in the world launch market.

Office of Commercial Space Transportation
800 Independence Ave, SW
Washington, DC 20591
(202) 267-8602

FEDERAL COMMUNICATIONS COMMISSION

The FCC is responsible for developing and administering policies and procedures concerning the regulation of telecommunications facilities and services under its jurisdiction and licensing of satellite and radiocommunication activities. In addition, the FCC represents the U.S. in international negotiations for satellite frequency allocations. Most of the employment opportunities in the office are in the area of law and licensing.

FCC International Bureau
2000 M Street, NW, Suite 800
Washington, DC 20554
(202) 418-0200

NATIONAL SCIENCE FOUNDATION

NSF is a federal agency whose aim is to promote and advance scientific progress. NSF funds research and education in science through grants and contracts to more than 2,000 colleges, universities, and research institutions. The agency operates no laboratories itself, but it could be a source of financing for research or scholarships.

National Science Foundation
4201 Wilson Blvd.
Arlington, VA 22230
(703) 306-1234

U.S. GEOLOGICAL SERVICE

The USGS manages, uses, and archives remotely sensed geologic data. The service has research programs in the areas of geologic mapping, tectonism, volcanism, global climate change, desert studies, exploration geology, impact crater studies, asteroid and comet radiometry, and image processing.

U.S. Geological Survey
2255 N. Gemini Drive
Flagstaff, AZ 86001
(520) 556-7018

EMPLOYMENT WITHIN NASA AND THE FEDERAL GOVERNMENT

At this time, opportunity for employment with NASA is quite limited. Downsizing requirements, budget limitations and restructuring have forced NASA centers to severely restrict permanent hiring opportunities to only the most critical vacancies. Alternatively, a number of NASA centers are filling vacant positions on a term or temporary basis. The list of NASA job opportunities should be reviewed carefully to determine which vacancies are open for consideration of non-NASA employees, and which vacancies are to be filled on a term or temporary basis.

How to Apply for a Federal Job

In the past, applying for a federal position required the completion of a standardized Government application form, Standard Form (SF) 171. Subsequent to the National Performance Review, the Office of Personnel Management (OPM) eliminated the SF-171. In its place, OPM made available an Optional Federal Job Application, but indicated that resumes or other application formats were acceptable for entry into the Federal Government. Since agencies may require a particular application format for current federal employees or agency applicants, you should review the vacancy announcement to insure that the proper format is being submitted.

To find out which positions are available, you should contact the agency or NASA field center that you are interested in working for or review the job listings posted on USA Jobs, the Office of Personnel Management's (OPM) web site providing a consolidated listing of Federal job opportunities.
http://www.usajobs.opm.gov

Federal opportunities within NASA can also be found at:
http://www.hq.nasa.gov/office/codef.codefp/jobinfo.html
http://huminfo.arc.nasa.gov/NASAvacancy.html

Another resource for available government positions can be found at:
http://www.fedworld.gov/jobs/jobsearch.html

Fedworld's site gathers employment opportunities from the official government personnel requests submitted to the Office of Personnel Management. The site allows for electronic searching of positions by keyword and location.

If you are unable to find a position within the government that meets your background and experience but would still like to work on a NASA-supported project, consider working for a private entity with a contract from NASA. Much of the work that is performed in support of NASA projects is done by contractors. For instance, the contract for the oversight and manufacturing of much of the Space Station, resides in a contract with The Boeing Company's Space and Defense Group.

MILITARY SPACE SECTOR

The military use of space preceded the establishment of NASA in 1958. However, it was not until the National Space Policy of 1978 that the military perspective and use of space was emphasized publicly as national policy. From the earliest days of the space program, the military has had a vital interest in the development of space systems. In the 1940s and 1950s, the military led the effort to develop and use the V2 rocket technology, developed by Germany for use in WWII, in U.S. missiles and rockets. In fact, the early rockets used by NASA for human spaceflight were derived from military missile designs.

Military uses of space include the utilization of satellites for:

- ▲ communications
- ▲ remote sensing of the environment for terrain modeling, meteorology, and oceanography
- ▲ reconnaissance and surveillance
- ▲ position and navigation determination
- ▲ early warning of missile launches

In addition, the military takes an active role in:

- ▲ launch operations
- ▲ ground station operations
- ▲ new technology research and development

The Department of Defense's efforts are coordinated by several agencies. The Pentagon is responsible for developing an overall architecture for the space needs of the various military operations. Space Command, a joint military organization comprised of the Air Force, Army, and Navy and headquartered in Colorado Springs, Colorado, is responsible for the overall maintenance and operations of the military's space assets. These assets include satellites and their launching, ground facilities and launch pads for the rockets, and ground networks to monitor the satellites and receive data from them. Each of the component commands —Air Force, Army, Navy, etc.—also has operations devoted to the utilization of space assets. These operations largely depend on the mission of the command. For example, the Navy is interested in the use of remotely sensed data for oceanography and determining sea conditions, while the Army has an extensive interest in using satellites to track troop movement and equipment on the ground.

Other agencies within the military or intelligence communities that are involved with utilizing, performing research in, or developing space assets include the National Reconnaissance Office, the National Security Agency, the Ballistic Missile Defense Organization, and the Central Intelligence Agency.

Finding a Space Job in the Military

Positions within the military sector are occupied not only by military personnel but by civilian personnel directly employed by the military or by civilian contractors. For active military personnel, available positions can be found by through normal command channels. Others interested in opportunities in the military should contact their local recruiting office. Civilian opportunities can be found via the same sources listed under civil government opportunities, in particular USA Jobs, the Office of Personnel Management's consolidated listing of federal job opportunities located at:

http://www.usajobs.opm.gov

You can also contact the agency directly at the below addresses:

Department of Defense—U.S. Space Command
250 South Peterson Blvd., Suite 116
Colorado Springs, CO 80914
(719) 554-3001

Department of Defense—The Pentagon
Washington, DC 20301
(703) 545-6700

ACADEMIC SECTOR

As would be expected, universities and non-profit research institutions are primarily involved with research. Study topics span the spectrum from the effect of microgravity on materials or biomedical drugs to new methods for improving propulsion efficiencies to evaluating sensor data on the chemical analysis of stars.

In addition to pure research, many academic organizations also develop instruments and other hardware to support their research endeavors. Research positions within the academic sector are usually reserved for faculty and university staff and therefore often require a Ph.D.

Research within this sector is primarily funded by government sources, such as NASA, the Department of Defense, and the National Science Foundation, or by private industry. NASA maintains a mission to fund research and annually awards more than $600 million dollars to universities and non-profit research institutions.

Some of the organizations in the Academic Sector can be found in Chapter 7: Colleges and Universities

At a large, diversified organization it is usually easier to transfer into your area of interest, after a year of working there, than to find the exact position you want immediately out of school. Once inside, you are more familiar with the organization and upcoming opportunities, and they are more familiar with your capabilities. Organizations want to keep good employees and openings are usually offered to current employees first.

Working in the Space Industry

BIOGRAPHY #4

Name: Paul Freid

Organization: U.S. Department of Defense

Current Profession: Space Systems Engineer

Degrees/School

B.S. Electrical Engineering, University of Lowell

Activities/Job Function

I have been working in a department whose goal is to develop secure communications techniques for satellite communication networks. These techniques are used in military, civilian and commercial communications systems. This requires knowledge of electronics, software, cryptography, contracting, and —most important of all—our customer's needs.

Why Space?

I wanted to work on leading edge technology, and early in my career I was given the opportunity to work in the space arena. The government agency that I am employed by allows me to utilize technologies years before their commercial potential or application can be realized.

Words of Wisdom

Choosing to work for the Department of Defense is one of the many options that may be available to you. If you want to work on the cutting edge of high-technology research and development, working with or for the government may provide you with that opportunity.

OPPORTUNITIES WITHIN THE SPACE INDUSTRY

Who Makes It Work?

"Software engineering is probably the most important skill to have to work in the space industry. It doesn't mean everything else is not important, but software with knowledge of other science and engineering disciplines is a critical skill in very high demand."
—Tracey Staley, University Relations Manager, Lockheed Martin

The following classifications are meant to provide a brief overview of the types of positions that are available within the space industry. Be forewarned. Different organizations will call jobs by different names. Additionally, there are many jobs out there that do not fit into the below categories. This section is only meant to give you a taste of the variety of jobs and tasks.

Cross Reference Table: Educational Degree vs Industry Opportunities

	Satellite Manufacturing	Ground Segment	Instrument Design	Launch Vehicles	Remote Sensing	Microgravity	Space Science	Manned Flight
Electrical Engineering	✔	✔	✔	✔		✔	✔	✔
Mechanical Engineering	✔	✔	✔	✔		✔	✔	✔
Propulsion	✔			✔			✔	✔
Power	✔							✔
Control Systems	✔	✔		✔		✔		✔
Software	✔	✔		✔	✔		✔	✔
Information Technology	✔	✔		✔	✔		✔	✔
Chemical Engineering				✔		✔		
Materials and Material Science			✔	✔		✔	✔	
Thermal Control	✔		✔	✔		✔		✔
Optics			✔		✔			
Structures	✔			✔				✔
Systems Integration	✔	✔		✔		✔		✔
Biology			✔			✔	✔	✔
Astrophysics			✔				✔	

TELECOMMUNICATIONS

Engineers specializing in this field work toward system optimizations. They develop ways to transmit and receive more information in less time and over greater distances; mobile systems that use less power and reduce the size and weight of ground equipment; and networks to transmit data seamless across satellite, ground, and wireless networks.

Sample Telecommunications Positions

Systems Engineer—Network Modeling
Develop and maintain telephony, paging traffic, and capacity models for worldwide communications network.

Radio Frequency Systems Engineer
Design, develop, and test satellite-based communications systems for data, video, and audio; perform satellite interference and RFI analysis; conduct satellite link analysis; perform site surveys; and develop systems integration documentation for field implementation.

Communications Systems Engineer
Perform various assignments in the following areas: messaging protocols (TCP/IP socket and RPC HDLC and Async serial); digital communications systems design (including DSP design, RF modem performance measurement and testing); real-time software applications; and feedback control systems.

Network Technicians
Requires technical experience in the operation and maintenance of electronic communication equipment. The technician will operate satellite and terrestrial communication systems; analyze, troubleshoot and resolve technical problems; operate communications test equipment; and provide technical support to customers and maintenance personnel.

SATELLITE DESIGN AND MANUFACTURING

Employees in this area use electronics and/or computers to design, develop, manufacture, and test the various systems that make up a satellite. Since the satellite will operate in an environment where repairs are difficult if not impossible to make, quality is an extremely important issue. Much of the hardware assembly and manufacturing takes place in a clean room to prevent dust and other particles from contaminating the system.

Technical positions include those that relate to the following:

- ▲ Bus Design
- ▲ Structure Design and Manufacturing
- ▲ System Integration
- ▲ Sensor Design
- ▲ Satellite Power Systems
- ▲ On-Board Software and Hardware
- ▲ Quality Control

The Importance of Satellites

Satellites are the primary driver of the space industry today, providing a means to generate or transmit data and information across great distances. From orbits above the Earth, satellites:

- ▲ Transmit voice, data, and facsimiles across the globe
- ▲ Enable mobile telecommunications across vast regions of the Earth
- ▲ Provide remote areas of the world with the infrastructure to access telecommunications and the Internet at lower cost
- ▲ Transmit and relay visual images for television and cable programming distribution, medical imaging and telemedicine capabilities, and tele-education services including distance learning
- ▲ Monitor and predict the Earth's weather
- ▲ Monitor the Earth's environment

How Satellites Work?

The primary components within a satellite are the:

- ▲ Bus—The electronic brain of the satellite, which processes and monitors all functions of the satellite
- ▲ Body—The physical structure, which holds or supports the other components.
- ▲ Power Supply—For Earth orbiting satellites, this generally consists of solar cells in an array and a storage device such as a battery. Other types of power supplies include nuclear devices (usually for interplanetary missions) or fuel cells for storage.
- ▲ Antennas and Associated Electronic Equipment—To receive, amplify, and transmit information
- ▲ Sensor Package—For remote sensing applications this could include devices to measure solar radiation, imagery of the Earth surface, or the local temperature of the ocean's surface.

Sample Classified Ads and Position Descriptions for Satellite Manufacturing

Spacecraft Preliminary Design Engineer
Electrical Engineering position involved with spacecraft design and fabrication and spacecraft electronics systems design. Responsible for spacecraft preliminary designs, mission payload and spacecraft electronics design and tradeoff analyses. Lead front-end spacecraft and ground segment design effort. Familiarity with spacecraft attitude control systems, and digital processor and telemetry/command subsystems and their interaction.

Satellite Digital Design Engineer
Design, test, and integration tasks. Some analog circuit design and spacecraft experience desirable. Proficient in designs utilizing microprocessors, associated memory, I/O and controller circuitry.

Jr. Digital Design Engineer
Perform spacecraft design, test, and integration tasks focusing on real-time design in satellite command, ranging and telemetry subsystems.

Mission Planning Engineer
Participate in spacecraft satellite operations planning and development of the spacecraft control center software and operational hardware while supporting mission operations and related activities such as prelaunch integration, postlaunch checkout and deployment of satellites.

Systems Engineer
Spacecraft systems and systems design with experience in both electrical and mechanical subsystems, electric power, propulsion, ADACS, thermal, and ground subsystems through integration and test.

Principal Mechanical Engineer
Responsible for the design, fabrication, integration, and testing of spacecraft structural and mechanical components & electromechanical systems and components. Experience in design verification, document production and analysis, cost analysis, design, test and manufacturing support and planning.

Power Engineer
Responsible for the design of electric power conversion and control equipment, including analysis, circuit design, electrical layout, prototyping, production support, test and system integration.

Systems Engineer
Familiarity with all spacecraft electronics subsystems and communications payloads required. Experience with preliminary design and analyses, advanced concepts, proven as well as advanced technology, and hands-on experience at the satellite/payload component design, fabrication, test and integration level.

Lead Engineer
Geosynchronous attitude determination and controls including experience in attitude control, spin axis attitude determination, analysis and simulation, and flight software specification/test support.

Satellite RF/Microwave Engineer
Geostationary satellite experience with design and development of C-band, S-band and other microwave frequency satellite communications hardware from design through integration. Proficient in communications analysis and circuit design and analysis in the S through Ku band, part selection, space qualification test, and subsystem and system integration and test.

Senior Engineering Technician
Responsible for building test fixtures, testing spacecraft components, and supporting engineering by assisting in the fabrication of bread boards. Field experience preferred. Ability to read and understand schematic diagrams and troubleshoot analog and digital circuits.

Flight Software Engineers
Develop onboard navigation, control, and communications software systems. Programming embedded in 80X86 based computers with SCCs and other common devices using ASM and /C++ under the VRTX OS.

Manager, Satellite Payload Engineer
Manage payload design, development, integration, and test.

Electronics Technicians
Integrate and test components for satellite ground stations and gateway centers.

GROUND SEGMENT OPERATIONS—SATELLITES

Ground segment positions include the following activities:

- ▲ monitoring satellite operations and the real-time analysis of the health of the satellite
- ▲ operating ground station hardware for processing of data
- ▲ developing software for planning the satellite and evaluating onboard performance and operations

▲ designing, developing, manufacturing, and testing electronic components that make up the ground station, including the antennas, relays, amplifiers, etc.

▲ manufacturing handheld or mobile transmitters and receivers for remote communications

▲ developing data storage and archive retrieval hardware and software

Use of Satellite Ground Stations

The ground station is used to receive data that originates from a satellite. Two types of data are received. The first is used for stationkeeping of the satellite to determine if it is operating within expected parameters. The second is the data that the satellite has been sent up for —telephone calls, remotely-sensed imagery, etc.

Remotely-sensed data, which is transmitted from the satellite, is stored on disk and recorded by time index and any other variables that are determined to be of necessity. Communications data is sent from the ground station to a local network where the data can be sent via the ground or through another satellite to its intended destination.

Ground Station Components

The major components of a ground station include the antennas that receive the data from space, software to analyze and interpret the data, and data storage devices such as tape drives.

In addition, numerous electronics devices, such as modems and amplifiers, are used to receive and process the data and enhance signal strength.

Sample Classified Ads and Position Descriptions for Ground Segment Personnel

Satellite Control Operators
Responsible for the daily operation of the ground system required to support satellite control functions. Duties include: routine satellite and ground station monitoring and control, response to satellite and ground system anomalies, and collection reporting of all satellite and ground station activities. Experience in any of the following is desired: satellite communication terminal operation/maintenance, communication network operation, or computer operations.

Mission Control Operators
These individuals will operate the Mission Control systems for configuring payload, maintaining the quality of the satellite broadcast, and controlling all access to the satellite by feeder link earth stations. The controllers will be responsible for the restoration of broadcast services in the event of satellite failures or impairment in the space segment.

Measurement and Instrumentation
Develop systems to measure and record physical phenomena and information to control environments and processes by means of various types of instrumentation, e.g. electrical, electronic, mechanical, and combinations. This work includes tracking systems, telemetry, radio, optical, and mechanical systems and subsystems. Included in this group are: sensors and transducers, heat and light measurement, measurement standards and calibration, automated control systems, and the electronics of materials.

Software Engineer—Ground Station
Develop UNIX workstation spacecraft command and control systems. Individuals will develop real-time command and control services, graphical user interface display services, orbit and attitude applications. Strong experience in C/C++. Orbital analysis requires experience with orbital mechanics and simulation of dynamic systems.

VSAT Implementation Specialists
Requires knowledge of VSAT hardware and systems, RF and IF systems, building construction practices, and VSAT installation types and procedures.

VSAT Applications Engineer
Familiarity with data communications protocols: TCP/IP, X.25, and SDLC.

Antenna Engineers
Design, fabrication, and test of UHF and microwave antenna systems. Knowledge of phase-tracked spiral antennas for interferometer arrays desired. Knowledge of analytical tools to support antenna design and installation interactions, including method of moments and geometrical theory of diffraction-based codes.

Communications Test Engineer
Compose test procedures, configure test hardware and perform tests on RF and communications equipment for spacecraft and launch vehicles. Knowledge of communications theory principles and experience with digital data transmission systems characterization.

Astrodynamics—Design and calculate the formulas and perturbations that determine a satellite's trajectory or orbit. Requires extensive mathematics and computer skills.

GROUND OPERATIONS—LAUNCH FACILITIES

Launch vehicles are operated and tested at specialized facilities which contain the necessary support infrastructure. In addition to the launch pad, the site generally contains a ground station area to monitor the launch, a payload processing facility where the payloads are mated to the fairing in the launch vehicle, and any number of storage tanks containing fuel, oxygen, and oxidizers. Launch sites in North America include government and commercial facilities at or near Kennedy Space Center in Florida, Vandenberg Air Force Base in California, and Wallops Flight Facility in Virginia; and commercial operations at sites in Alaska and in Manitoba, Canada. Many of the employees work in areas related to safety, test, ground control operations, electrical and mechanical support, and construction and facility design.

LAUNCH VEHICLES

Most of the labor in the launch vehicle industry is related to the design, development, manufacture, and test of the vehicle. Positions range from technicians in a machine shop or assembly floor to engineers designing subsystems, solving assembly problems, or overseeing test operations. As an employee, you might be involved in one of the following areas:

- ▲ Manufacturing—Components & Structures
- ▲ Component & Vehicle Testing
- ▲ Quality Control
- ▲ Engineering—Materials, Propulsion, Structures, Control Systems, Thermal Analysis

- ▲ Drafting/CAD Design
- ▲ Aerodynamics
- ▲ Propellant Testing & Development

Launch Vehicle Components

A launch vehicle is made up of several major pieces of hardware—

- ▲ Engine
- ▲ Fuel
- ▲ Support Structure including fuel and/or oxidizer tanks
- ▲ Fairing

Of these, the fairing is probably the least familiar term, but it is one of the most critical pieces of hardware. The fairing is the part of the rocket that attaches the payload to the launch vehicle and allows it to release the payload into the proper orbit. Part of the fairing is a separation device that allows this to occur. If the fairing does not operate as designed, then the payload might be damaged or left in a useless orbit. In both cases, the mission would be a failure.

Sample Classified Ads and Position Descriptions for Launch Vehicle Personnel

Thermal Engineer
Develop large thermal models of space vehicles using SINDA/TRASYS or PC_ITAS.

Propulsion Engineer
Participate in the design, development, integration, test and evaluation of propulsion-related activities. Background in cryogenic and/or hyperbolic propulsion systems. Strong background in propulsion component (valves and regulators) design for turbo machinery.

Propulsion Systems Engineer
Opportunities exist in the conversion of energy into power for space systems. Specializations include liquid propulsion systems, solid propulsion systems, electrical propulsion and power, energy conversion, nuclear energy processes, nuclear propulsion and power, chemical energy processes, internal flow dynamics, and propulsion system dynamics.

Vehicle System Engineer
From mission planning through launch operations, the Integration and Test Operations Group has a variety of exciting opportunities for engineers with hands-on hardware experience. Requires excellent organizational skills and the ability to work in a technical team.

Draftsperson/Computer Aided Design
Draw sketches and schematics of systems, subsystems and hardware to specifications provided by the engineers.

Experimental Facilities and Equipment
Includes the design, development, test, evaluation, operation, and management of aerospace research and development facilities and equipment for experimental and operational purposes. Positions include the following specialties: launch and flight operations, experimental tooling and equipment, fluid and flow dynamics, electrical experimental equipment, and experimental facilities techniques.

Materials and Structures
Positions involved with research, design, development, test, and evaluation (RDDT&E) of aerospace vehicle structures and the study of their behavior in flight regimes. These positions are also engaged in research into the behavior and characteristics of materials for use in flight vehicles and systems. Included in this group are the following specialties: materials, structural mechanics, aerospace metals, basic properties of materials, polymers, refractory compounds, friction and lubrication, structural mechanics, flight structures. Majors include ceramics or ceramic engineering, metallurgy or metallurgical engineering, physics, engineering (various), and chemistry.

Fluid and Flight Mechanics
Positions concerned with the study and investigation of dynamics of aerospace vehicle flight and the establishment of criteria for aerospace vehicle design based on the dynamics of flight. Positions may also be concerned with the investigation of the interaction of the vehicle in flight and the environment. Positions are also engaged with research, development, design, test and evaluation of systems to guide and control the vehicle in flight. Specializations in this group include flight

mechanics, control and guidance systems, fluid mechanics, magnetofluid dynamics, aerostructural dynamics, vehicle acoustics, heat transfer, stability, control and performance, flight vehicle atmosphere environment, and basis properties of gases.

Flight Systems
Positions in this specialty are concerned with systems integration, reliability studies, evaluation of systems and subsystems design and performance characteristics. Included in this specialty are positions performing research, development and evaluation of manufacturing and quality assurance programs. Following specialties are included in this group: reliability, flight systems test, experimental manufacturing techniques, quality assurance, electrical systems, and piloted space flight systems.

HUMAN SPACEFLIGHT ACTIVITIES

Positions involved with human spaceflight have historically been driven by NASA. Activities involve developing control devices and hardware to keep the crew alive and functioning. In addition to the above hardware, government-sponsored research is performed on methods to evaluate and improve human conditions in space with respect to the physical and psychological problems faced by crews during long missions.

MISSION HARDWARE

The most important hardware and software associated with manned flight is the life support systems for the vehicles as well as those for extravehicular activities. While the most familiar piece of hardware is the spacesuit, systems to control oxygen levels and waste removal are extremely vital. Other hardware used in manned flight applications relate to cooking, bathing, and cleaning. Most of the new devices are funded either by NASA or the Russian Space Agency—the two organizations that place humans in orbit.

MEDICAL AND BIOTECHNOLOGY

- ▲ Medical Research—What is the impact of gravity on life and living systems? How does the human body adapt in space? What can we learn about space travelers that could be applied to conditions on Earth? (i.e. solving the problem of bone marrow loss by astronauts might lead to a cure for osteoporosis.)

- ▲ Life Studies—Theoretical and experimental research on the effects of space environmental stresses upon living organisms and systems. Specialties include: biochemical processes, psychological studies, plant studies, physiological studies, molecular biodynamics, radiobiological studies, and neurobiology.

- ▲ Man-Machine Systems—Theoretical and experimental research on the effects of space environment stresses upon man functioning as an integral component of a man-machine system for flight and exploration. Specialties include: physiology, human performance studies, environmental control, manned systems engineering, and bionics studies.

REMOTE SENSING

Remote sensing is simply the ability to collect information from a distant location. Using satellites that contain sensors that are able to provide us with information about the Earth, users on the ground are able to analyze this data for a variety of purposes.

In general, remote sensing organizations perform at least one of following types of activities:

- ▲ develop the systems and sensors that will collect the data
- ▲ develop software to analyze the data
- ▲ develop software or reports that will convey the data in an understandable format
- ▲ perform research and analysis of the data for earth monitoring, intelligence gathering, or geographical information systems (GIS)

Once the data is collected and relayed to the ground, the task of understanding the data is set to begin. Using graphical software packages, analysts are able to combine different data sets to generate a picture that will interest the end-user. For instance, in order to assist a civil planner in locating the best site for a new facility, the following data sets might need to be combined:

- ▲ a digital terrain model showing the height and slope of the ground
- ▲ a population distribution model
- ▲ a data set showing roads and power lines
- ▲ data showing underground streams

Analysts interpreting remotely sensed data have always been in high demand. With the development of automated software allowing the data to be more easily processed as well as the increase of data being derived from the sensors, the number of these positions is anticipated to increase over the next decade.

New software applications and analysis products are enabling the remote sensing industry to target customers not normally associated with the space industry—farmers, architects and regional planners, geologists, and even archaeologists.

In general, many positions within the industry are involved with:

- ▲ data interpretation and analysis
- ▲ development of graphic interfaces
- ▲ computer simulations
- ▲ computer programming

These positions can be found in all three space communities—commercial, civil, and military—and include the analysis of data gathered for military intelligence, environmental monitoring and prediction, and weather monitoring and prediction.

Sample Classified Ads and Position Descriptions for Remote Sensing and Photogrammetry

Imagery & Geospatial Analysts and Engineers
These positions are involved with software engineering; imagery analysis and exploitation; imagery tasking, collection and dissemination; geospatial/mapping analysis; cartography; photogrammetry; hydrography and communications; image processing; and multi- and hyperspectral imagery requirements and training.

Sensor Designers
Requires scientists and engineers with experience in any or all of the following: optics, RF and microwave design, and electronics.

Software Engineers—Graphical Information Systems
Mid-sized image handling systems to include image exploitation, image database applications and simulations. Will work with non-technical experts to develop accurate simulations using customer application software.

Senior Analyst
Position involves experience with digital photogrammetric workstations, stereo imagery control, and feature analysis.

Remote Sensing Team Leader
Ability to program and handle operating systems and natural resources data in a GIS environment.

Photogrammetrists and Technicians
Experience in aerotriangulation, CAD, digital orthophoto, stereo compilation, and systems management.

SPACE SCIENCE AND ASTROPHYSICS

Almost all of these positions are research oriented and are located at universities or at government research facilities.

▲ Space Sciences—Includes positions engaged in the study and investigation of atmospheres and space phenomena, the heavenly bodies and their characteristics, astrophysics, and celestial mechanics.

The following specializations are included in this group: Aeronomy, ionosphere, fields and particles, stellar studies, lunar and planetary studies, meteoroid studies, solar studies. College majors for the above include physics, astronomy, meteorology, geology, geophysics, astrophysics, or other appropriate fields of basic physical science. Courses of interest include electronics, optics, materials, vibration, high-vacuum theory, heat transfer, and aerospace instrumentation.

▲ Astrophysics—The primary aim of the astrophysicist is to understand the physical processes governing the behavior of the atmospheres and space environments of the Earth, Sun, and other planetary bodies. It includes research on the Earth's thermosphere, ionosphere, and magnetosphere as well as the solar wind and coupling between these regions. Other areas where astrophysicists get involved include activities related to galactic events, cosmic rays, solar wind, background radiation, pulsar research, black hole accretion, globular clusters, and the detection of new stars, galaxies, and planets.

▲ Astrobiology—Theoretical and experimental research on the nature and origin of life in the universe. Involves studies directed toward understanding the nature and basic mechanics involved with the synthesis of biologically significant compounds, the evolution and adaptation of life forms, and the development of life detection systems and devices suitable for space flights and exploration. Specialties include chemical evolution, biological adaptation, and life-detection systems.

GLOBAL POSITIONING SYSTEM (GPS)

The GPS is an all weather, continuous operation, space-based radio navigation system.

Positions in this industry sector are predominantly in the manufacture and design of electronic receivers and in marketing the products to new consumers. Shortly after the turn of the century, GPS devices will be used not only to provide directional information for aircraft and automobiles but also to enable the automated and precision use of machinery, such as that used to plow large farms. In addition, a consumer market linking GPS receivers to mobile telephones will enable a host of new companies to provide services in vehicle and valuable tracking and directional and emergency assistance to backpackers, hikers, and boaters. These services are anticipated to create numerous jobs for sales and marketing personnel as well as electronic technicians.

MICROGRAVITY

The use of the microgravity environment for research is a nascent commercial market which is expected to grow significantly in the next several years. Most microgravity research is currently performed within the government or academic communities. Hardware is developed at both government and commercial organizations. Opportunities exist within firms for:

- ▲ designing and developing experiment hardware
- ▲ analyzing data derived from the experiments
- ▲ marketing microgravity hardware and facilities to potential users and researchers
- ▲ researching how microgravity affects processes that are well known on earth. The mixing and properties of materials, fluid flow, and crystal growth, all function much differently in a microgravity environment.

Researchers within the academic and research community generally have a specialty background in areas such as molecular biology, biochemistry, cell-to-cell interaction, crystallography, or materials.

SOFTWARE AND INFORMATION TECHNOLOGY

Almost any software or information technology skill that exists in the broader economy can be found in the space industry. Skilled persons are needed in network management, database design and development, object oriented techniques, software integration and test, software architecture and design, scientific/analysis software, flight software development, and spacecraft and ground systems software development.

Sample Classified Ads

Software Engineer
Develop reusable real-time embedded software for satellite systems. Perform detailed design, coding and testing of software units. Support software subsystem integration. Requires a B.S. degree in Computer Science, Math, Physics, Electrical Engineering, or related field and knowledge of C or C++.

Software Engineer
Basic duties involve attitude control system design, analysis, and simulation. Tasks include working for senior engineers in the areas of spacecraft attitude control, control system stability analysis, control law design, system hardware component modeling and analysis, and system-level performance analysis. Knowledge of C programming language and computer-aided simulation software.

Software Engineer
Conduct analyses and designs in one of the following orbit, mission, and system analysis disciplines: ascent trajectory generation, launch system performance, orbital transfer, constellation coverage, multi-body relative motion, coverage, orbit ephemeris generation and determination, orbit perturbation modeling, mission operations, and reentry.

Software Engineer, Sr.
Perform science data analysis. Work closely with scientists in data reduction and analysis. Maintain instrument operations software including data maintenance, distribution of science data, database management, and operations management.

OTHER TECHNICAL POSITIONS

- ▲ Quality Assurance Technician or Engineer—All systems that must function in space require that the risk due to faulty hardware or software be kept at an absolute minimum (many subsystems require 99.999% reliability or better). Material tolerances are to sub-fractions of an inch. People in these positions are responsible for establishing and maintaining a quality assurance program, including subcontract/vendor control, incoming inspection, in-process inspection, process controls, integration/test support, qualification/ acceptance test definitions, and database development.

- ▲ Analyst—Consulting positions that provide guidance and expertise to a client, usually to determine the best possible solution or the effect of a suggested solution.

- ▲ Systems Architecture Studies—A limited number of positions within the industry are available for developing the requirements necessary for achieving a goal, such as the necessary space systems for military operations and long-range planning related to lunar bases and Mars missions. Many of these positions are in the military or at NASA and universities.

BUSINESS AND ADMINISTRATIVE POSITIONS

Numerous professional careers that do not require degrees in engineering and science are available in the space industry. Some of these are positions responsible for the following:

Marketing and Sales
Policy Analysis
Management Consulting
Publishing
Administration (support)
Procurement Management
Public Affairs
Legal and Licensing
Advertising Sales
Business Analysts
Finance
Insurance
Accounting
Reporting
Contract Administration
Human Resources
Technical Writing and Editing
Government Relations
Customer Service Manager
Proposal Coordinator

Sample Non-Technical Classified Ads and Position Descriptions

Management Consulting
Assist companies in identifying market opportunities and developing strategies for competing in today's marketplace.

Program Management
Positions involve the technical and management direction of programs or projects. Incumbents of these positions prepare technical plans, budget and cost estimates, determine resources required for programs and projects, and schedule phases of the work.

Director of Business Development
Identify new business opportunities, coordinate opportunity evaluation and bid decisions, provide interface with customers, identify pertinent new markets for our products.

Technical Writer
Write and edit sections for proposals, coordinate proposal tasks, verify compliance with specifications, and proofread corporate literature.

Proposal Coordinator
Duties include developing and executing proposal production and coordination plans, establishing and maintaining libraries of resumes, request for proposals, and submitted proposals. A familiarity with software programs such as Word, Pagemaker, and Excel is required.

When thinking about elective courses to take or what minor to pursue, consider expanding your knowledge of computers and software programming. It could prove to be very valuable in your technical career. You should also consider courses related to business and management, as the skills and perspective they offer may complement your background.

Working in the Space Industry

BIOGRAPHY #5

Name: Les Johnson

Company: National Aeronautics and Space Administration, Marshall Space Flight Center

Title/Profession: Principal Investigator and Project Manager

Responsibilities

I am the Principal Investigator of the Propulsive Small Expendable Deployer System (ProSEDS) mission and the study manager for the Multiple Tethered Satellites for Ionospheric Studies mission. In addition, I am responsible for several of NASA's concept definition efforts for future space missions using tethered satellites. The ProSEDS mission will demonstrate electrodynamic tether propulsion of small spacecraft in low earth orbit when it flies in 2000.

Degrees/School

B.A. Chemistry and Physics, Transylvania University
M.S. Physics, Vanderbilt University
Graduate, International Space University

Career Path

I planned from a very early age (pre-high school) to work for NASA or in the space industry. I studied chemistry and physics, later emphasizing space physics in graduate school, to facilitate my career. Unfortunately, upon earning my Masters Degree, NASA was not hiring and I took a job near NASA MSFC working for General Research Corporation. While at GRC I made several professional contacts within NASA and was hired there 3 1/2 years later. I now work in the Program Development Directorate, which is our future projects office. Prior to my current tether-related activities, I led the design and definition of the $80M Magnetosphere Imager spacecraft and mission for NASA's Office of Space Science.

Why Space?

I have always been fascinated by space and, fortunately, I have the technical skills that enable me to work in the field.

Words of Wisdom

Be tenacious and don't give up! It took more than three years for me to get a job at NASA. I have been told on numerous occasions that I am successful in large part due to my optimism and 'can-do' attitude. This formula can work for others as it has for me.

7 COLLEGES & UNIVERSITIES

"The greatest gain from space travel consists in the extension of our knowledge. In a hundred years this newly won knowledge will pay huge and unexpected dividends."—Wernher von Braun

GETTING STARTED: UNIVERSITY PROGRAMS

Choosing a college for your undergraduate or graduate studies can be a time-consuming and agonizing task. In addition to balancing location, course offerings, school reputation, professors' backgrounds, and financial aspects, you have also chosen the option of trying to find a school that will allow you to pursue a career in the space industry. Keep in mind that a space career does not necessitate a specific discipline or even that you participate in space-related programs while gaining your education.

Take for instance a technical career. Many engineers and scientists in the industry have a degree(s) in any one of the following: Electrical Engineering (circuits, control systems, etc.); Mechanical Engineering (structures, propulsion, fluid mechanics, etc.); Materials and Materials Engineering (composite materials, ceramics, high-temperature metals, etc.); Physics, Chemical Engineering, Software and Computer Science Engineering; Robotics; etc.

Opportunities within the space industry are as broad as the degrees you may pursue—from engineering and science to business and management, policy analysis, and sales. The space industry has a position for almost anyone.

Even if you choose a school that does not have a major program or research effort in space, chances are that there are at least one or two professors with a similar interest or a small research study involved with disciplines that can be applied to space. Remember that your time at a university is meant to be used to gain a broad education. Your time in industry will be spent applying that knowledge.

An Author's Experience: Finding Space Activities at a University Without a Devoted Curriculum

Our example is R.P.I. (Renssellaer Polytechnic Institute) located in upstate NY, near the Vermont and Massachusetts border. During the 1980s, while one of the book's authors attended RPI, the school did not have a specific program or numerous course offerings devoted to space or space sciences. A student approaching the school would have been told about degrees offered in aeronautical engineering. As time went on and Mr. Sacknoff learned more about the space industry, he found several programs and activities at the university. The President of the University at the time was George M. Low, a former administrator of NASA. There was a professor with a research grant to study how space-based lasers could be used in earth-to-orbit propulsion systems. Additionally, there were major efforts in advanced materials and robotics, for which the departments were receiving funding from NASA to study microgravity materials processing and robotic systems.

UNIVERSITIES WITH SPACE-RELATED PROGRAMS

Several universities maintain specialized programs related to space. The list on the opposite page has been compiled only to show the breadth of programs and activities available. It is not meant to provide a complete list of programs. You will find —at the end of this chapter —a more in-depth list of schools, along with information on how to contact them.

Arizona State University	Astrophysics, Satellite Design
Auburn	Space Power Systems
Baylor University	Astrophysics
Boston University	Space Physics
Carnegie Mellon University	Robotics
Clarkson University	Microgravity Materials Processing
George Washington University	Policy
Kansas State University	Biology
New Mexico State University	Telecommunications and Telemetering
North Carolina State University	Mars Mission Research
Ohio State University	Remote Sensing
Pennsylvania State University	Propulsion, Communications, Space Sciences, Biotechnology
Rensselaer Polytechnic Institute	Materials, Microgravity, Robotics, Laser Propulsion
Rutgers University	Remote Sensing and GIS
South Dakota State University	Remote Sensing and GIS
Stanford	Astrophysics and Space Physics
Texas A&M	Space Power, Electromagnetics and Microwaves
University of Alabama	Microgravity—Biotechnology and Materials Processing
University of Alaska	Space Physics, Meteorology
University of Chicago	Astronomy and Astrophysics
University of Colorado—Boulder	Telecommunications, Space Structures and Controls, Remote Sensing, Microgravity
University of Florida	Solar Power, Nuclear Power
University of Houston	Microgravity Materials Processing
University of Idaho	Microelectronics
University of North Dakota	Policy

UNDERGRADUATE PROGRAMS

In choosing an undergraduate program, we recommend that you treat your education as a broader opportunity rather than focus on one specific school because it performs space research. Attending a school with a specific program may give you a clearer understanding of a specific part of the industry, but as an undergraduate, establishing a broad background may be more important in the long run. Keep in mind that a large number of people change or modify their career paths as time goes on. What you think you want to do today, may not be what you want to do tomorrow.

TYPES OF SPACE-RELATED COURSES

(Titles taken from courses catalogs of RPI, University of Colorado, and University of North Dakota)

TECHNICAL COURSES

Space Vehicle Design
Space Science and Exploration
Earth System Science
Introduction to Orbital Mechanics
Asteroids, Meteorites, & Comets
Satellite Information Processing
Wireless and Cellular Communications
Observational Astronomy
Computer Vision Image
Combustion Systems Introduction
Spaceflight Dynamics
Telecommunications Theory and Applications
Propagation Effects on Satellite and Deep Space Telecommunications
Aerospace Vehicles and Facilities Operations
Life Support Systems
Observational Astronomy
Global Change
Human Factors in Space
Technical Issues in Space
Celestial Mechanics
Quasars and Cosmology
Astrophysics I, II
Theory of Propulsion
General Manufacturing Processes

BUSINESS AND POLICY

Strategic Implications of Space
Space Treaties and Legislation
Remote Sensing Policy & Law
International Telecommunications Policy
Strategic Planning for Telecommunications
Fundamentals of Marketing Technology
Space Policy and International Implications
Soviet/Russian Space Program
Engineering Economics
Introduction to Space Technology
Strategic Implications of Space

GRADUATE PROGRAMS

A graduate degree is meant to focus your early career in a given direction—either by emphasizing a more narrow specialization or by adding a discipline to your desired goal (MBA in addition to a B.S. Engineering).

Two of the most important considerations in choosing a graduate program are your advisor and your research project. As with undergraduate activities, even schools without a narrow degree program may have a professor who can provide the opportunity for you to do research in a specific area. The U.S. government offers hundreds of millions of dollars in research grants to professors and schools throughout the nation. In addition, private industry sponsors numerous research endeavors. To find the graduate program and professor you want, you will need to undertake a research effort to determine your options. You may choose to make this task as simple as asking your undergraduate advisor to suggest a list of programs to investigate or as complicated as calling all potential universities and inquiring with whom to talk.

NETWORKING TO FINDING GRADUATE OPPORTUNITIES

Before calling schools, unless you know exactly where you want to go, it is highly recommended that you talk with some people in industry who are working in the specific area or field that you are hoping to pursue. It is more than likely that they may be able to tell you of an opportunity that you may not find otherwise.

There are many ways to find expert advice. The most direct way would be to ask your alumni relations department, your undergraduate department chairman, or even alumni at your fraternity/sorority. Contacts are everywhere; you only need to ask.

A list of some university programs placing an emphasis on space can be found at the end of this chapter.

ENHANCING YOUR EDUCATION—GETTING EXPERIENCE TO INCREASE YOUR EMPLOYMENT PROSPECTS

Regardless of what your major is, you can increase your chances of being offered a position and improve your resume by participating in research projects or by going on a cooperative education (COOP) assignment.

Companies place great emphasis on practical knowledge as compared to book knowledge. The ability to work with a team, knowing how to handle problems that arise on a project, and the ability to organize the tasks needed to finish a project, are skills that are better learned by doing than in a classroom.

COOPERATIVE EDUCATION PROGRAMS

COOP affords undergraduate students the opportunity to take a six-month to one-year leave of absence from school to work at a private or public institution and gain real world experience. These programs are usually coordinated by the Placement or Career Counseling Department at your university. It is their role to arrange for employers to interview on campus. If you decide to pursue the COOP option and would like to find an assignment with a space focus, it is recommended that you do some research on your own. Many of the larger space companies offer COOP programs. As it is highly unlikely that they will all visit your campus for interviews, you should plan to contact them on your own or with the help of your COOP office.

ON-CAMPUS RESEARCH ACTIVITIES

Another option for gaining real-world project experience is through supporting a professor's sponsored research project. Although most research opportunities are reserved for graduate students, many professors save a few slots for undergraduates. While larger schools or those with extensive research budgets will inevitably have more opportunities, there will also be more competition for the prime activities.

Other opportunities exist at those universities that house major non-profit research facilities. Several such examples are the Charles Stark Draper Laboratory with MIT, the Applied Physics Laboratory at Johns Hopkins University, the Space Dynamics Laboratory at Utah State University, and the Institute for Earth, Oceans, and Space at the University of New Hampshire.

Students should also talk with individual professors to see if they have a need for support.

SUMMER RESEARCH OPPORTUNITIES AT NASA

NASA sponsors a variety of research opportunities during the summer, both at the field centers themselves and in conjunction with other institutions.

The Space Life Sciences Training Program, co-sponsored by and located at Florida A&M University, offers undergraduate college students the opportunity to participate in an intensive six-week training course, during which they will experience a complete overview of space life sciences.

Florida A&M—College of Pharmacy
106 Honor House
Tallahassee, FL 32307
(904) 599-3636

The NASA Academy is a Goddard Space Flight Center program designed to introduce university students to the laboratories of the Center. The 10 week session involves approximately 24 students annually. Three days a week the students assemble in plenary sessions and deal with a unique subject ranging from spacecraft design to the latest findings from the Mission to Planet Earth or the Hubble Space Telescope. Sessions also include discussions of policy issues, such as finance and proposal evaluation, and off-site visits to contractors and NASA facilities.

NASA Academy
Goddard Space Flight Center
Office of University Programs
Greenbelt, MD 20771
(301) 286-2000

Additionally, each NASA Field Center provides the opportunity for a limited number of graduate and undergraduate students to support ongoing research activities. Students who are interested in these opportunities need to contact the Manager of University Programs at the NASA Field Center at which they are interested in working.

FINANCING YOUR EDUCATION—FELLOWSHIPS AND SCHOLARSHIPS

Many of the space-related fellowships and scholarships that exist are offered by industry associations, NASA, or other government agencies. When looking for financing, it is recommended that you investigate all sources including the industry associations mentioned in chapter 9. A number of fellowships and scholarships are listed below:

Dr. Robert Goddard Scholarship
The National Space Club awards a $10,000 scholarship in memory of Dr. Robert Goddard, America's rocket pioneer. The award is given to stimulate the interests of talented students to advance scientific knowledge through space research and exploration. The award is open to U.S. citizens in at least their junior year at an accredited university.

National Space Club
655 15th Street, NW
Washington, DC 20005
(202) 639-4210

Global Change Research Graduate Student Fellowship
The goal of this NASA program is to train the next generation of PhD's in Earth science and engineering. Students conduct research in the areas of climate and hydrologic systems, ecologic systems and dynamics, solid Earth processes, solar influence, human interactions, and data and information systems. Selected students will receive a one-year fellowship of $20,000, renewable for up to three years.

NASA Headquarters
300 E Street, SW
Washington, DC 20546
(202) 358-0000

Graduate Student Researchers Program

Fellowships of up to $22,000 are awarded for 1 year and are renewable for up to three years to graduate students whose research interests are compatible with NASA's programs in space science and aerospace technology. A majority of the awards are in the fields of astrophysics, earth science, life sciences, solar system exploration, space physics, and microgravity science and applications.

NASA Headquarters
300 E Street, SW
Washington, DC 20546
(202) 358-0000

National Defense Science and Engineering Graduate Fellowships

The Department of Defense offers approximately 90 three-year graduate fellowships in a variety of research disciplines. Stipends range from $16,500—$18,500 annually.

NDSEG Fellowship Program
200 Park Drive, Suite 211
PO Box 13444 Research Triangle Park 27709
(919) 549-8505 fax: (919) 549-8205
http://www.battelle.org/ndseg/ndseg.html

Small Satellite Scholarship Program

Offered annually at the AIAA/Utah State University Small Satellite Conference for the student(s) with the best application/research technology project that would benefit the small satellite industry. Scholarships have historically ranged from $1,000—$10,000.

Utah State University
Space Dynamics Laboratory
Logan, UT 84322-9700
(801) 755-4287

Space Grant Consortia

Located in all fifty states and Puerto Rico, the Space Grant program is designed to support space-related educational activities within each state. As part of its mission, the program offers many undergraduate and graduate scholarships. Contact NASA Headquarters or visit their web site to get the contact points in your state.

NASA Headquarters
300 E Street, SW
Washington, DC 20546
(202) 358-0000
http://deimos.ucsd.edu/space-grant/NASAspacegrant.html

Dr. Theodore von Karman Graduate Scholarship Program

Awards up to ten scholarship s of $5,000 to graduating Air Force ROTC seniors pursuing graduate degrees in the fields of science, mathematics or engineering prior to active duty.

Aerospace Education Foundation
Lee Highway
Arlington VA 22209-1198
(703) 247-5839

ASPRS Awards Programs

The ASPRS (American Society for Photogrammetry and Remote Sensing) administers a number of scholarship and award programs for undergraduate and graduate students interested in remote sensing and photogrammetry.

ASPRS
5410 Grosvenor Lane, Suite 210
Bethesda, MD 20814
(301) 493-0290

Zonta International Amelia Earhart Fellowship Awards

Awards approximately 35 scholarships of $6,000 each to women who hold a B.S. in science or engineering and have completed one year of graduate school.

Zonta International
557 West Randolph Street
Chicago, IL 60661
(312) 930-5848

Air Force Laboratory Graduate Fellowship Program
For doctoral study in the areas of science and engineering that are of interest to the Air Force, the fellowship pays for tuition and fees plus a stipend for up to three years.

USAF
c/o SE Center for Electrical Engineering Education
1100 Massachusetts Avenue
St. Cloud, FL 34769-3733
(407) 892-6146

AFCEA Education Foundation Fellowship
Provides up to $25,000 to students pursuing masters or doctorate degrees in the areas of electrical, electronic, or communications engineering, physics, or computer science.

AFCEA
4400 Fair Lakes Court, Fairfax VA 22033
(800) 336-4583 x6149

Some University Programs with an Emphasis on Space

State	University/College
AK	University of Alaska Geophysical Institute 903 N Koyukyuk Drive, Fairbanks, AK 99775 (907) 474-7954 Environmental Research
AL	Auburn University Space Power Institute 231 Leach Center, Auburn, AL 36849 (334) 844-5894 Power Systems/Electronics
AL	University of Alabama at Birmingham Center for Macromolecular Crystallography 1918 University Blvd. THT-Box 79 UAB Station, Birmingham, AL 35294-0005 (205) 934-5329 Microgravity/Biotechnology

AL	University of Alabama Consortium for Materials Development in Space 301 Sparkman Drive, Research Institute Building Room M65, Huntsville, AL 35899 (205) 890-6620 Microgravity
AL	University of Alabama Center for Microgravity & Materials Research M-65 RI Building, Huntsville, AL 35899 (205) 890-6050
AZ	Arizona State University Mechanical and Aerospace Engineering Box 876106, Tempe, AZ 85287-6106 (602) 965-2823
AZ	Arizona State University Center for Meteorite Studies Box 872504, Tempe, AZ 85287-2504 (603) 965-6511
AZ	Arizona State University Space Photography Laboratory Dept. of Geology, Box 1404, Tempe, AZ 85287-1404 (602) 965-7029
AZ	University of Arizona Digital Image Analysis Laboratory Dept. of Electrical & Computer Eng., Tuscon, AZ 85721 (520) 621-2706
AZ	University of Arizona Lunar & Planetary Laboratory 1629 East University Blvd. Space Sciences Building #92, Tucson, AZ 85721 (520) 621-6963
AZ	University of Arizona Space Engineering Research Center 4717 East Fort Lowell, Tuscon, AZ 85712 (520) 322-2304
CA	Naval Postgraduate School Space Systems Academic Group—Code SP 777 Dyer Rd, Room 200, Monterey, CA 93943-5110 (408) 656-2948

CA	Stanford University Center for Science & Astrophysics Varian 302F, Stanford, CA 94305 (650) 723-1439
CA	Stanford University Communication Satellite Planning Center Bldg 360, Room 361, Stanford, CA 94305-4053 (650) 723-3471
CA	Stanford University Space Telecommunications and Radio Science STARLAB/Electrical Engineering Department 226 Durand Bldg., Stanford, CA 94305-9515 (650) 723-8121
CA	Stanford University Satellite Systems Development Laboratory Dept. of Aeronautics & Astronautics, Durand 269 Stanford, CA 94305 (650) 723-8651
CA	University of California—Berkeley Space Sciences Laboratory Berkeley, CA 94720 (510) 642-0561
CA	University of California—San Diego California Space Institute/Scripps Institution of Oceanography 9500 Gilman Drive, La Jolla, CA 92093-0221 (619) 534-2830
CA	University of California Santa Barbara Institute for Computational Earth System Science 6832 Ellison Hall, Santa Barbara, CA 93106 (805) 893-4885
CO	University of Colorado—Boulder Center for Aerospace Structures Campus Box 429, Boulder, CO 80309-0432 (303) 492-6838
CO	University of Colorado—Boulder Center for Space Construction Campus Box 529, Boulder, CO 80309-0529 (303) 492-2556

CO	University of Colorado—Boulder Center for the Study of Earth from Space Campus Box 216, Room 318, Boulder, CO 80309 (303) 492-5086
CO	University of Colorado—Boulder Center for Low-G Fluid Mechanics & Transport Phenomena Dept. of Chemical Eng. Campus Box 424, Boulder, CO 80309-0424 (303) 492-7517
CO	University of Colorado—Boulder Interdisciplinary Telecommunications Program ECOT317, Campus Box 530, Boulder, CO 80309-0530 (303) 492-8916
CO	US Air Force Academy Astronautics Department 2354 Fairchild Drive, Suite 6J71 USAF Academy, CO 80840-6224 (719) 333-1110
CO	Vet Colorado Engineering Research Center OT3-27, Campus Box 0424, Boulder, CO 80309-0409 (303) 492-3633
CT	Yale University Astronomy Department 260 Whitney Ave., New Haven, CT 06511 (203) 432-3000
DC	Catholic University of America Robotics and Control Laboratory School of Engineering, 620 Michigan Ave, NE Washington, DC 20064 (202) 319-4787
DC	George Washington University Space Policy Institute 2013 G St, NW, Suite 201, Washington, DC 20052 (202) 994-7292
DC	International Space University 700 13th Street, NW, Suite 950, Washington, DC 20005 (202) 237-1987

DE	University of Delaware Center for Composite Materials Composites Manufacturing Science Lab Newark, DE 19716-3144 (302) 831-8149
FL	Embry-Riddle Aeronautical University Space Education, Research & Technology 600 South Clyde Morris Blvd. Daytona Beach, FL 32114-3900 (904) 226-6474
FL	Florida Institute of Technology School of Extended Graduate Studies 150 West University Blvd., Melbourne, FL 32901 (407) 768-8000
FL	University of Florida Innovative Nuclear Space Power & Propulsion P.O. Box 116502, Gainesville, FL 32611-6502 (352) 392-1427
FL	University of Florida Solar Energy and Energy Conversion Lab Dept. of Mechanical Engineering, PO Box 116300 Gainesville, FL 32611 (352) 392-0812
FL	University of Miami Remote Sensing Laboratory P.O. Box 248003, Coral Gables, FL 33124 (305) 284-3881
GA	Georgia Institute of Technology School of Earth and Atmospheric Sciences 221 Bobby Dodd Way Atlanta, GA 30332-0340 (404) 894-3893
ID	University of Idaho Microelectronics Research and Communications Institute Moscow, ID 83844-1024 (208) 885-6500

IL	University of Chicago Department of Astronomy and Astrophysics 5640 South Ellis Ave., Chicago, IL 60637 (773) 702-1234
IN	Purdue University Laboratory for Applications of Remote Sensing 1202 Potter Building, Room 220 West Lafayette, IN 47907-1202 (317) 494-6305
KS	Kansas State University Center for Gravitational Studies in Cellular and Development Biology Ackert Hall, Manhattan, KS 66506-4901 (913) 532-6615
KS	University of Kansas Space Technology Center 2291 Irving Hall Drive Raymond Nichols Hall, Lawrence, KS 66045 (913) 864-4775
MA	Boston University Center for Remote Sensing 725 Commonwealth Ave., Boston, MA 02215 (617) 353-9709
MA	Boston University Center for Space Physics 725 Commonwealth Ave., Boston, MA 02215 (617) 353-5990
MA	MIT Lincoln Laboratory 244 Wood Street, Lexington, MA 02173-9108 (617) 981-5500
MA	MIT Center for Space Research 70 Vassar Street, Cambridge, MA 02139-4307 (617) 253-1456
MA	MIT Man-Vehicle Laboratory 77 Massachusetts Ave., RM 37-219, Cambridge, MA 02139 (617) 253-7805

MA	MIT Space Engineering Research Center Aeronautics and Astronautics Department 77 Massachusetts Ave., Bldg. 33, Room 207 Cambridge, MA 02139 (617) 253-7510
MD	Johns Hopkins University, The Applied Physics Laboratory/Space Department 100 Johns Hopkins Road, Laurel, MD 20723-6099 (301) 953-5000
MD	University of Maryland Center for Satellite & Hybrid Communication A.V. Williams Bldg., College Park, MD 20742 (301) 405-7900
MD	University of Maryland East-West Space Science Center Department of Physics, College Park, MD 20742-3280 (301) 405-8052
MD	University of Maryland Space Systems Laboratory Dept. of Physics, College Park, MD 20742 (301) 405-7353
MI	University of Michigan Atmospheric Ocean & Space Sciences Space Research Bldg., 2455 Hayward Street Ann Arbor, MI 48109-2143 (313) 936-0493
MI	University of Michigan Center for Space Terahertz Technology 3228 EECS Bldg., 1301 Beal Ave., Ann Arbor, MI 48109-2122 (313) 764-0501
MI	University of Michigan Space Physics Research Laboratory 2455 Hayward Street, Ann Arbor, MI 48109-2143 (313) 936-7775
MN	University of Minnesota Space Physics Research Group, School of Physics & Astronomy 116 Church Street, SE, Minneapolis, MN 55455 (612) 624-8089

MO	Washington University Earth & Planetary Sensing Laboratory CB1169, 1 Brookings Drive, St. Louis, MO 63130 (314) 935-5610
MO	Washington University McDonnell Center for the Space Sciences 1 Brookings Drive, Campus Box 1105 St. Louis, MO 63130-4899 (314) 935-6257
NC	North Carolina State University Mars Mission Research Center 1009 Capability Drive, Box 7921, Raleigh, NC 27695-7921 (919) 515-5931
ND	University of North Dakota Department of Space Studies Corner of University and Campus Drive P.O. Box 9008, Grand Forks, ND 58202-9008 (701) 777-2480
NE	University of Nebraska Center for Advanced Land Management Information Technology 113 Nebraska Hall, Lincoln, NE 68588-0517 (402) 472-8197
NH	University of New Hampshire Institute for Earth, Oceans, and Space Space Science Center, 39 College Rd, Morse Hall Durham, NH 03824-3525 (603) 862-0322
NJ	Rutgers University/Cook College Center for Remote Sensing & Spatial Analysis College Farm Road, New Brunswick, NJ 08903-0231 (732) 932-9631
NM	New Mexico State University Communications Systems Simulation Laboratory Electrical & Computer Eng. Dept. Box 30001, Dept. 3-0 Las Cruces, NM 88003-0001 (505) 646-3115

NM	New Mexico State University Ctr. for Space Telemetering & Telecommunications Systems Box 30001, Dept. 3-0, Las Cruces, NM 88003-0001 (505) 646-3012
NM	NM Institute of Mining and Technology Astrophysics Research Center Department of Physics, Campus Station, Socorro, NM 87801 (505) 835-5328
NM	University of New Mexico Institute for Space Nuclear Power Studies Farris Engineering Center, Room 239 Albuquerque, NM 87131-1341 (505) 277-0446
NV	University of Nevada CIASTA 7010 Dandini Blvd., Reno, NV 89512 (702) 673-7312 Remote Sensing
NY	Clarkson University International Center for Gravity Material Science & Applications Box 5814, Potsdam, NY 13699-5814 (315) 268-7672
NY	Cornell University Center for Radiophysics & Space Research Space Sciences Building, Ithaca, NY 14853 (607) 255-4341
NY	Cornell University National Astronomy and Ionosphere Center Space Sciences Bldg., Ithaca, NY 14853-6801 (607) 255-3735
NY	Cornell University School of Electrical Engineering 224 Phillips Hall, Ithaca, NY 14853 (607) 255-4109
NY	Niagara University Space Settlement Studies Project Timon Hall, Niagara University, NY 14109 (716) 286-8094

NY	Rensselaer Polytechnic Institute Mechanical and Aerospace Dept. Troy, NY 12180-3590 (518) 276-6545
NY	Rensselaer Polytechnic Institute NY State Center for Advanced Technology in Automation CII Bldg., Room 8015 Troy, NY 12180-3590 (518) 276-8087
NY	Rensselaer Polytechnic Institute Materials Science & Engineering Department Materials Research Center, Room 14 Troy, NY 12180-3590 (518) 276-6000 Space Materials Processing
OH	Ohio State University Center for Mapping 1216 Kinnear Road, Columbus, OH 43212 (614) 292-1600
PA	Carnegie Mellon University Field Robotics Center 5000 Forbes Ave., Pittsburgh, PA 15213 (412) 268-6559
PA	Penn State University Communications & Space Sciences Laboratory 318 Electrical Engineering East University Park, PA 16802-2707 (814) 865-6337
PA	Pennsylvania State University Propulsion Engineering Research Center 106 Research Building East, Bigler Road University Park, PA 16802 (814) 863-6272
PA	Pennsylvania State University Center for Cell Research 117 Research Office Building University Park, PA 16802 (814) 865-2407

PA	Pennsylvania State University Office for Remote Sensing of Earth Resources Land & Water Research Bldg., University Park, PA 16802 (814) 865-9753
RI	Brown University Department of Geological Sciences Lincoln Field Bldg., Box 1846, Providence, RI 02912 (401) 863-3243
SD	South Dakota State University Office of Remote Sensing Engineering Research Center Harding Hall 228, Brookings, SD 57007-0199 (605) 688-4184
TN	University of Tennessee Space Institute B.H. Goether Pkwy, Tullahoma, TN 37388 (931) 393-7100
TN	Vanderbuilt University Box 1824, Station B, Nashville, TN 37235 (615) 322-2771
TX	Space Vacuum Epitaxy Center University of Houston Science & Research Bldg. 1 4800 Calhoun Road, Room 724 Houston, TX 77204-5507 (713) 743-3621
TX	Texas A&M University Center for Space Power 223 Weisenbaker ERC, College Station, TX 77843-3118 (409) 845-8768
TX	University of Texas at Austin Center for Space Research 3925 West Braker Lane, Suite 200, Austin, TX 78759 (512) 471-5573
TX	University of Texas at Dallas Center for Space Sciences P.O. Box 830688, M/S F022, Richardson, TX 75083-0688 (972) 883-2851

UT	University of Utah High Energy Astrophysics Institute 201 James Flecher Bldg., Salt Lake City, UT 84112 (801) 581-5505
UT	Utah State University Space Dynamics Laboratory 1695 North Research Park Way, North Logan, UT 84341 (435) 797-4600
UT	Weber State University Center for Aerospace Technology College of Applied Science & Technology 1805 University Circle, Ogden, UT 84408-1805 (801) 626-7272
VA	Old Dominion University Virginia Space Flight Center P.O. Box 6369, Norfolk, VA 23508 (757) 440-4020
VA	Virginia Polytechnic Institute Center for Wireless Telecommunications Bradley Dept. of Electrical Engineering 340 Whittemore Hall, Blacksburg, VA 24060 (540) 231-4461
WA	University of Washington Aerospace & Energetics Research Program Aerospace & Engineering Research Building—Room 120 Seattle, WA 98195-2250 (206) 543-6321
WI	University of Wisconsin—Madison Space Astronomy Laboratory 1150 University Ave., Madison, WI 53706 (608) 263-4680

WI	University of Wisconsin—Madison Space Science and Engineering Center 1225 West Dayton Street, Madison, WI 53706 (608) 263-6750
WI	University of Wisconsin—Madison Wisconsin Center for Space Automation and Robotics 1415 Engineering Drive, Room 2348, Madison, WI 53706 (608) 262-5524 Robotics, Automation, Space Manufacturing

People skills are very important. The stereotype of the lone engineer or scientist working in a secluded lab is mostly a myth. You will find it necessary to be able to interact with colleagues, supervisors, assistants, people in other departments, as well as customers and clients.

8 FINDING A JOB FOR THE FIRST TIME...OR THE NOT SO FIRST TIME

"The Earth is a cradle of the mind, but we cannot live forever in a cradle."—Konstantin Tsiolkovsky

How do you interview? How do you write a good resume? These are not topics that we plan to cover—there are hundreds of books written by experts on these subjects. This chapter and those that follow on networking will focus on the resources available within the space industry to aid you in your search for a career or a new job.

SPACE—JUST LIKE ANY OTHER INDUSTRY

After completing your search for employment in the space industry, you will realize that the resources and knowledge you uncovered during your search are similar to those used when searching for employment in other industries—classified advertisements, industry publications, the World Wide Web, etc.

Similarly, industry recruiters will give the same kind of advice to those considering employment in this business that they would to others.

- ▲ Today's global economy stresses flexibility, the ability to adapt skills to different tasks. Employees who can handle a wide range of tasks will be in higher demand.
- ▲ Keep in mind that even when companies downsize or reduce staff, many are still hiring to fill needs.

- ▲ Small and mid-size companies have historically grown faster than large companies and do much of the hiring in an industry.
- ▲ It is usually easier to transfer from within a company than to get the exact position you want when first applying.
- ▲ If you're on the outside of the industry looking in, it may not be as difficult as it appears to obtain a position within the space industry. Although the retail and space industries would appear to have little in common, an employee from the retail industry could bring in needed knowledge in distribution and marketing.
- ▲ Traits such as communication skills and the ability to work well in a group are highly desirable.

IDENTIFYING INFORMATION RESOURCES

When searching for publicized opportunities, there are several places to look:

- ▲ Classified Advertising
- ▲ Industry Publications
- ▲ Industry Associations
- ▲ World Wide Web
- ▲ Career Resource Libraries at Universities
- ▲ Public Libraries

CLASSIFIED ADVERTISING

While classified advertising in industry magazines has historically been limited, publications such as *Space News, Via Satellite, Aviation Week and Space Technology, Aerospace America, Satellite Communications,* and *Launchspace Magazine* do contain a handful of job openings. The titles of other industry publications can be found by going to your public library and asking the reference librarian for *Gale's Directory of Publications* or by reviewing an industry directory, such as the *United States Space Directory.*

Local papers in cities with a heavy industry concentration, such as the *Washington Post*, the *Los Angeles Times*, *Florida Today* and the *Houston Chronicle*, usually contain more ads than the industry publications. A search in a recent high-technology supplement of the *Washington Post* revealed more than 100 space-related job descriptions from around the United States.

Industry Magazines

Space News 6883 Commercial Drive Springfield, VA 22159 tel: (703) 658-8400 http://www.spacenews.com	**Via Satellite** 1201 Seven Locks Rd, Suite 300 Potomac, MD 20854 tel: (301) 340-1520 http://www.phillips.com
Aviation Week & Space Tech. 1200 G Street, NW Suite 922 Washington, DC 20005 tel: (202) 383-2300 http://www.awgnet.com	**Launchspace** 7929 Westpark Drive, Suite 100 McLean, VA 22102 tel: (703) 749-2324 http://www.launchspace.com
Satellite Communications 5660 Greenwood Plaza Blvd. Suite 350 Englewood, CO 80111 tel: (303) 220-0660	**Aerospace America** AIAA 1801 Alexander Bell Dr., Ste 500 Reston, VA 20191 tel: (703) 264-7500 http://www.aiaa.org
GPS World PO Box 10460 Eugene, OR 97440 tel: (541) 343-1200 http://www.gpsworld.com	**PE&RS (Remote Sensing)** 5410 Grosvenor Lane, Suite 210 Bethesda, MD 20814-2160 tel: (301) 493-0290

INDUSTRY ASSOCIATIONS

As part of their missions, most associations usually have a goal of educating members and the public. As part of this goal, many of them maintain a formal or informal mechanism for notifying members about open positions. Usually, the associations promote these in one of three ways: via advertisement in their member publication, via advertisement on their web site, or via job fairs or posted job bulletins at their conferences.

With the diversity of skills needs and the activities and skills taking place within the space industry, a wide range of associations exist focusing on everything from specific technical disciplines to enthusiasts interested in promoting the exploration of space. Joining an association allows you to keep on top of the industry in your area of interest and provides you with the opportunity to network and meet others at conferences and meetings.

A list of associations and regular industry conferences can be found in Chapter 9: Networking.

WORLD WIDE WEB

The World Wide Web is becoming the best source for finding available positions. Many companies, especially the larger ones, offer connections from their main homepage to a database of open positions within their company. In addition, there exist a number of WWW sites specifically devoted to employment opportunities. Some of these contain more information than others, and many do not contain categories directly related to space employment. On a broad employment site, you may have to be patient and type in a number of search terms. So, don't be discouraged if the first site you evaluate doesn't have any positions related to your keyword.

In our review of the WWW, we have identified the following sites, which you may want to explore:

Space Industry WWW Sites with Employment Opportunities

- ▲ http://www.spacebusiness.com
- ▲ http://www.aiaa.org/employment/index.html
- ▲ http://www.nab.org/ECLI

Employment classified sites focusing on general opportunities throughout all sectors of the economy include:

- ▲ http://www.careerpath.com (Career Path)
- ▲ http://safetynet.doleta.gov/jobbank.htm (America's Job Bank)
- ▲ http://www.cweb.com (Career Web)
- ▲ http://www.jobweb.org/catapult/igenlist.htm (Job Web)
- ▲ http://www.washingtonpost.com/wp-adv/classifieds/careerpost/front.htm

Among the many companies that use the WWW to inform prospective employees about opportunities and company policies are the following:

Lockheed Martin	http://www.lmco.com/jobs.html
Boeing	http://www.boeing.com or http://metis.bna.boeing.com/coredata/job/homecar.html
Orbital Sciences	http://www.orbital.com
Motorola	http://www.mot.com/Employment
Hughes	http://www.hughes.com/jobsnav.html

CAREER RESOURCE CENTERS AT COLLEGES AND UNIVERSITIES

University Career Placement Offices are another good source of job opportunities. While these offices are set up to help current students or graduates of the university, many allow limited use by the general public as a community service. Career placement offices receive open position notices from many companies that have had success in the past hiring students or graduates from that university. In addition to announcing entry level positions, many firms will also submit announcements for more senior personnel.

Most universities subscribe to an online system developed by Jobtrak. Allowing on-campus usage without a password and off-campus usage with a password, the system contains all the job postings that private organizations want to announce to members of that universities' community. Each company specifies which universities they want to have access to their specific announcements. One open position may be viewed only by Cornell and MIT while another position may specify Rensselaer Polytechnic Institute, Princeton, and the University of Maryland. Passwords and logon information can be received from the alumni or career placement offices at each university. The WWW site to access these listings can be found at:

http://www.jobtrak.com

PUBLIC LIBRARIES

Don't forget a search at your public library. In addition to numerous books on writing resumes, interview skills, etc., the reference section of the library contains numerous publications, which are generally out of the financial reach of the average job seeker. Among these resources will be:

- ▲ directories providing details on public, private, and non-profit organizations
- ▲ information on the major employers in a given geographic area

OTHER RESOURCES

Appendix A contains a list of organizations found within the current edition of the *United States Space Directory* and has been provided as a resource. Before writing to an organization, we recommend that you call first or visit their WWW site to learn more about the organization and to find the appropriate point of contact for recruiting/personnel matters.

Publications, such as the above directory, include information on an organization's specializations and capabilities. They also provide contact information such as the address, phone, fax, Internet site location, and key contacts within the firm.

Details on this publication and other reference publications can be found on our homepage at http://www.spacebusiness.com under the Bookstore option.

OTHER OPTIONS

Keep in mind that, in addition to direct employment, there are other options you may want to consider—specifically, temporary or contract assignments or starting your own firm.

TEMPORARY & CONTRACT EMPLOYMENT

If you are interested in evaluating an organization, evaluating an employment activity you are capable of performing but are unsure if you want a career in, or want to work short-term to see a different area of the world, you may want to investigate non-permanent employment. Also, if you are trying to work for a specific company that does not have a current opening, a temporary position can allow you to get your foot in the door. This approach allows companies to evaluate employees and their performance prior to hiring them on a permanent basis.

Contract employment positions are generally short-term assignments (3 months to two years) for technical positions. These positions typically pay substantially better than direct employment as they usually carry no benefits or any guarantee of having the contract extended for any period of time. Needless to say, there is no job security as a contract employee.

Additionally, many companies also make use of local temporary agencies to fill administrative, professional, or technical positions. Most of these agencies can be found in your local telephone book.

Remember: Although designed to be temporary, the job that you're hired to do could lead to a permanent position with the firm if those around you are satisfied with your performance. This is similar to some of the reasons why Cooperative Education (COOP) assignments and summer employment are important to finding future employment. (As mentioned in the Chapter on Colleges and Universities)

Temporary contract technical (space and non-space) positions throughout the country and the world can be found in *Contract Employment Weekly*, a national magazine devoted to this type of employment.

Contract Employment Weekly
11515 NE 118th Street
Kirkland, WA 98083
(425) 823-2222

STARTING YOUR OWN COMPANY

Got a better way of doing something? Perhaps there is a technology that you developed or worked with and think is great, but no one at your organization wants to do anything commercial with it. Hundreds of new firms in the industry are being created by visionaries who simply choose the option of stepping outside an organization. Many space industry ventures do not require billions of dollars to get off the ground. Office expenses for software and hardware have become so modest that it does not always require a vast sum to set up a business. If you're an entrepreneur and have some experience, perhaps starting your own company is the answer.

RESOURCES AVAILABLE TO START-UP ORGANIZATIONS

SBIR/STTR

The Small Business Innovation Research Program (SBIR) and Small Technology Transfer Program (STTR) are federal programs that were started as a means to ensure that small firms were able to get some of the vast amounts of research and development funding spent by the government. The SBIR competition was designed to set aside a pool of money for which only small companies could compete. Over time the program has grown to where almost $1 billion annually is given out by the eleven federal agencies. Among the agencies that have SBIR programs and space-related topics in their solicitations are NASA, the Department of Defense, and NOAA.

Companies who win (historically, about 1 in 7 or 8) can receive up to $100,000 for a Phase I—6-month effort, which can be followed by a Phase II award of up to $750,000. Companies retain the non-government intellectual property rights for four years.

For solicitations listing the technical topics under this program, contact the Small Business Administration or any of the participating federal agencies.

SBA 7A LOAN GUARANTEE PROGRAM

The Small Business Administration offers guarantees of up to 80% on loans made by banks or lending institutions to small businesses. Those SBA-backed loans allow the financial institution to make riskier loans to new companies without a track record and without substantial collateral. Contact for this program is through your local bank or through the regional SBA Office (check your local telephone book). Not all banks participate or are familiar with the program, so you will need to inquire.

MENTOR PROGRAMS

Many university business schools have established programs that can provide an individual or a company with a mentor— someone who has worked in or who has retired from industry. These mentors usually have extensive experience and provide advice on a variety of business-related topics, usually at no or minimal charge. Even if your local college or university doesn't have a program, inquire about whether anyone there is aware of someone who might be willing to assist you with advice.

OVERSEAS OPPORTUNITIES

While many in the United States would like to work overseas—in Europe perhaps—the opportunities outside the U.S. for an American citizen are limited—unless they have several years of technical experience in a needed discipline. To reduce unemployment, many countries have imposed strict rules and regulations that state how and when companies can hire someone from outside the country.

For instance, a French company must first offer the position to a French national and then to a member of the European Community before it can be offered to a citizen of the United States.

Note that the U.S. has similar laws. Historically, however, it has been much easier for a foreign national to get a job in the U.S. than the other way around.

If you are determined to find employment opportunities overseas, we recommend trying these methods:

1. Working as an employee of a U.S.-based company assigned to an overseas office
2. Working with a contract employment agency that specializes in bringing in high-tech talent for short-term (usually up to two years) assignments
3. Contacting non-U.S. companies and government agencies directly to determine their needs and ask for their assistance
4. Searching the Internet

TECHNICAL CONTRACT EMPLOYMENT ORGANIZATIONS

Two companies specializing in bringing technical talent from the United States overseas are Design Team International and Howard Organisation.

Design Team Inc.
100 W Beaver Creek Rd
Richmond Hill, Ontario L4B 1H4 Canada
tel: (905) 707-8480
fax: (905) 707-8483

Howard Organisation Ltd.
Project House
110/113 Tottenham Court
London, Great Britain W1A 1BX
tel: +44 (171) 388.4555
fax: +44 (171) 388.0455

SPACE ORGANIZATIONS LOCATED OUTSIDE THE UNITED STATES

The following WWW sites provide you access to information on a number of non-U.S. organizations and government agencies.

Canadian Space Agency	http://www.space.gc.ca
European Space Agency	http://www.esrin.esa.it
NASDA (Japan)	http://nasda.go.jp
Int'l Space Science Institute	http://ubeclu.unibe.ch/issi/index.html
Italian Space Agency	http://www.asi.it
British National Space Center	http://www,open.gov.uk/bnsc/bnschome.html
CNES (France)	http://www.cnes.fr

Working in the Space Industry

BIOGRAPHY #6

Name: Max Stolack

Organization: United Technologies Corporation

Current Profession: Project Engineer, Launch Vehicles

Degrees
B.S. Mechanical Engineering

Activities/Job Function
Our group is responsible for coordinating the activities of several departments involved with the manufacture, assembly, and test of experimental jet and rocket engines. One of our main responsibilities is troubleshooting problems so that the engine is built and tested properly, and is on-time and on-budget.

Why Space?
The thing I like best about the space industry is that you can see the big picture. When we take our engine to the test range and watch it work, we can see the results of our labor.

Words of Wisdom
Do not be afraid of changing jobs or careers. Space is made up of a whole range of organizations, big and small, private and government. Each person will find a different level of happiness in a different type of organization.

9 NETWORKING: YOUR KEY TO SUCCESS

NETWORKING EXAMPLE
Prior to moving to Washington DC, one of the authors met with an industry consultant. The consultant didn't know of any positions within his organization, but he suggested that I contact three other people he knew and said to mention that he referred me. One of those three suggested talking to a small company down the hall, which was expanding at the time. This led to a job offer.

Regardless of your current or future needs, networking is one of the most important activities for advancing your career. In many cases, it is not what you know, but whom you know, and it's whom you know that can give you information few have. Statistics show that classified ads make up only 10-15% of all available jobs.

To network effectively, think about who might be a good contact. Your list should include the following: family, friends, college professors, college alumni, present and former supervisors, people you've met at industry events, and people you've met at social events.

Keep in mind—

1. Contacts provide valuable information about existing or upcoming opportunities, which are never published.
2. A contact may not be helpful immediately, but may become valuable in the future. Example: A recent graduate met an insurance broker at a luncheon. Three years later, the broker started a venture capital fund and

invested in an idea the former student had. Remember: You never know who will end up where.

3. A company recruiter is more likely to hire someone who comes with a recommendation from a current employee. Networking gives you contacts on the inside.

QUESTIONS TO ASK

You will need to tailor your questions according to the contact you are talking to—someone working in your area of interest, personnel manager, friend of the family, etc. However, the following questions are meant to help guide you in genera questions you may want answered.

- ▲ What does your company look for when hiring?
- ▲ What sort of positions is your company filling today or expect to be filling in the future?
- ▲ How easy/difficult is it for me to find employment in my area of interest?
- ▲ What knowledge and skills are the most important?
- ▲ How important is it for a person in a position of this type to have a knowledge of the industry? (e.g. This would be more important for a job in public relations, media, legal, etc. than for the duties of a propulsion engineer.)
- ▲ What training and experience do you think would be valuable?
- ▲ How often do you feel bored or frustrated in your type of work?
- ▲ What do you do on a typical day?
- ▲ What sort of practical experience would be valuable? Knowledge of...?
- ▲ What previous jobs led to your position?
- ▲ Are there similar types of positions in the area that I am interested in?
- ▲ What are the required qualifications and training for an entry-level position?
- ▲ Can you recommend any courses to take?
- ▲ What related fields or studies do you suggest I explore?
- ▲ Do you have any special advice you can offer?

Most importantly, always ask as your last question...
I appreciate your time; is there anyone else that you suggest that I talk with?

In turn, expect that the people you talk with will ask you the following questions:

- ▲ Where did you get my name? (It's usually helpful to mention this in advance, as they may be more willing to meet with you, if they know who referred you.)
- ▲ Why did you get interested in the industry?
- ▲ What courses have you taken? What is your background?
- ▲ What are your short/long-term career goals?

PUT WHAT YOU'VE LEARNED IN A DATABASE

In order to keep track of the details of your networking experience, it is recommended that you take advantage of today's technology to keep notes on the following. Your records should prove useful as well as providing you with discipline for the future.

- ▲ Name and title of each person you contacted
- ▲ The company and address
- ▲ How did you find this person? Who referred you?
- ▲ The date of contact.
- ▲ Your method of contact—Via telephone, letter, at a meeting
- ▲ The context of your discussion
- ▲ Any follow-up that is required.

WHERE TO NETWORK

Now that you've determined what questions to ask and what information you seek, it is time to get involved and seek out people with similar interests. Much of the networking in an industry takes place at the following:

- ▲ Sessions at conferences & symposia
- ▲ Exhibitions at conferences
- ▲ Association events including industry speakers and networking breakfasts and luncheons

CONFERENCES, SYMPOSIA, EXHIBITIONS

Conferences and their exhibit areas are usually sponsored by industry publications, associations, or technical societies. If you don't have the money to attend, don't worry. You can usually gain admittance to the exhibit area for free or a nominal charge.

Walking around the exhibit floor is a highly valuable networking exercise. In fact, if your primary purpose is networking, rather than expanding your understanding of a subject being explained by a speaker or presenter, the exhibit floor is where you want to be. Instead of sitting in a "classroom", you are talking to people.

This does not mean you should avoid the sessions. If you are registered for the conference, and there is a speaker talking on the area in which you are interested in working, it is highly advised that you listen to the talk and approach the speaker after the session has ended. Coffee breaks are often the best opportunities to network.

PROFESSIONAL ORGANIZATIONS

American Institute of Aeronautics and Astronautics

Founded in 1930, the AIAA has more than 60 years of serving the needs of the aerospace industry. Among the services AIAA provides to the aerospace profession are six scholarly journals, two technical book series, an aerospace library, a database containing more than 1.9 million abstracts, more than 20 technical meetings each year, at least six exhibitions, the monthly magazine *Aerospace America*, and more than 90 technical and standards committees.

AIAA also maintains local chapters for members to meet and network, as well as student chapters at a number of university campuses.

Among AIAA's conferences and exhibits are: Global Air & Space (Washington DC—April/May); Joint Propulsion Conference (Various Locations—July); Space Programs and Technologies (Huntsville AL—September); and Aerospace Sciences (Reno, NV—January). A complete list can be obtained by calling AIAA or by reviewing their web site.

AIAA
1801 Alexander Bell Drive, Suite 500
Reston, VA 22091
(703) 264-7500
http://www.aiaa.org

American Astronautical Society

Founded in 1954, the AAS works with scientific and technical members dedicated to the advancement of space science and exploration in the U.S. In addition to producing a number of publications, the AAS holds three major meetings annually, including the Goddard Space Policy Symposium (in March), an annual conference (various locations), and a meeting focusing on Military space activities.

American Astronautical Society
6352 Rolling Mill Place, Suite 102
Alexandria, VA 22152
(703) 866-0020

Society of Satellite Professionals International (SSPI)

The goal of the organization is to promote professional development in the field of satellite applications, increase public awareness, and create a global network of contacts and associates.

2200 Wilson Blvd., Suite 102-258
Arlington, VA 22201
(703) 243-8948

National Space Club

Located in Washington D.C, with chapters in Huntsville, AL and Los Angeles, CA, the National Space Club offers monthly luncheons with industry speakers and sponsors the Goddard Dinner, a black-tie affair. The Club is geared towards industry executives.

2000 L Street, NW
Suite 710
Washington, DC 20036
(202) 973-8661

American Society for Photogrammetry and Remote Sensing (ASPRS)

The ASPRS is a technical organization open to firms and individuals involved with photogrammetry, remote sensing, photo interpretation, GIS, and their applications to archaeology, military reconnaissance, urban planning, meteorological observation, forestry, agriculture, construction, and topographic mapping. The Society publishes a monthly magazine for its members on technology and developments of interest and sponsors or co-sponsors a number of technical conferences and symposia.

ASPRS
5410 Grosvenor Lane, Suite 210
Bethesda, MD 20814
(301) 493-0290
http://www.asprs.org/asprs

PUBLIC INTEREST AND OTHER GROUPS

National Space Society

The mission of the NSS is to provide a forum for public participation in the space program and help insure the program is responsive to public priorities, inspires space enthusiasts, and promotes space research, exploration, development, and habitation. The NSS has over 100 regional chapters throughout the country and in several countries overseas. It publishes *Ad Astra* ("..To the Stars") magazine for its members and holds an annual space development conference. In addition, local and regional chapters sponsor their own events. Information on local chapters can be found on the NSS WWW site or by calling the National Headquarters.

National Space Society
600 Pennsylvania Ave, SE, Suite 201
Washington, DC 20003
(202) 543-1900
http://www.nss.org

The Planetary Society

The Planetary Society was founded in 1980 to support the exploration of the solar system and to continue the search for extraterrestrial life. With approximately 100,000 members in 100 countries, it is the largest space interest group in the world.

The Planetary Society
65 N Catalina Avenue
Pasadena, CA 91106
(818) 793-5100
http://planetary.org/tps

U.S. Space Foundation

The Foundation, many of whose members represent industry and government, works through educational programs to develop a better understanding and awareness of the beneficial uses of space and an increase in student motivation. The Foundation conducts graduate courses for teachers, tries to incorporate space education into school curricula, and supports research as it relates to space. The Foundation also sponsors the annual National Space Symposium.

U.S. Space Foundation
2860 South Circle Drive
Suite 2301
Colorado Springs, CO 80906
(719) 576-8000
http://www.inovatec.com/ussf

INDUSTRY CONFERENCES

In addition to the conferences mentioned above as part of the activities of the associations, a number of other conferences occur including the following:

Satellite 'XX

Via Satellite magazine's annual satellite conference is the premier satellite meeting in the industry and offers technical sessions, symposia, and a large exhibit area devoted primarily to organizations involved with satellite and component manufacturing and the utilization of satellites for telecommunications and other applications. The conference is held annually in Washington D.C.

Via Satellite
1201 Seven Locks Rd, Suite 300
Potomac, MD 20854
(301) 340-1520

National Space Symposium
Sponsored by the U.S. Space Foundation and held annually in the spring in Colorado Springs, Colorado, the conference and exhibit feature a range of companies and organizations involved with the military, civil, and commercial space programs. Recent themes of the conference have focused on the commercial business and policy issues of the industry.

U.S. Space Foundation
2860 South Circle Drive, Suite 2301
Colorado Springs, CO 80906
(719) 576-8000

Utah Small Satellite Conference and Exhibit
This annual September show is devoted to the technical and business aspects of small satellites.

Utah State University
Conference and Institute Division
Logan, UT 84322-5005
(801) 797-0423

Paris Air Show
Held bi-annually during the summer, alternating with the Farnborough Air Show in England, the show is mostly related to organizations interested in civil and military aircraft. With more than 200,000 attendees, the Paris Air Show also can boast that it has the largest attendance of people interested in space in the world. As mentioned earlier in this book, many space companies are also involved with aviation and/or defense, and job seekers should not rule out working in an aviation or defense department and transferring to a space-related position later.

Department of Commerce
International Trade Administration
Office of Aerospace Room 2128
Washington, DC 20230
(202) 482-2232

OTHER PROFESSIONAL GROUPS HOLDING NETWORKING BREAKFASTS, LUNCHEONS, SPEAKERS

Women In Aerospace
PO Box 44492
Washington, DC 20026-4492
(202) 547-9451

Space Transportation Association
2800 Shirlington Rd, Suite 405
Arlington, VA 22206
(703) 671-4116

Washington Space Business Roundtable
655 15th Street NW, Suite 300
Washington, DC 20005
(202) 639-4105

Florida Space Business Roundtable
P.O. Box 273
Cape Canaveral, FL 32903
(407) 868-6983

BUSINESS-ORIENTED ASSOCIATIONS

International Space Business Council
2055 North Fifteenth Street, Suite 300
Arlington, VA 22201
(703) 524-2766
http://www.spacebusiness.com

10 ARE YOU ASTRONAUT MATERIAL?

"For most, working in space is a dream not a career."

While we realize that attaining astronaut status is a life long dream for many, becoming an astronaut is far more difficult than being picked for the National Basketball Association™ or the National Football League™. This fact is not meant to discourage, but it is reality. In future decades, we may all have the chance to travel into space. For now, opening up the space frontier remains the domain of a select cadre of people.

DREAMING OF SPACE

No doubt, many of you reading this book aspire to become 21st century versions of the "Right Stuff". That was the expression coined by author Tom Wolfe in his book and movie of the same name. Wolfe's characterization of tough-as-nails experimental aircraft test pilots and the fearless set of America's first astronauts for the Mercury program created a persona that is still in evidence today.

But the astronaut corps has come a long way since the early 1960s when Alan Shepard was heralded as the first U.S. astronaut. He was shot into the air and plopped into the Atlantic Ocean in May 1961, rocketing through air and space for all of 15 minutes. While historic, that suborbital jaunt pales in comparison to later space sojourns.

Today, some 35 years later, more than 300 people have headed for orbit, the majority of them boosted there courtesy of a U.S. Space Shuttle. Fifteen groups of astronauts have been chosen between 1959 and 1995. Early in the next century, astronauts will take up semi-permanent residence in the International Space Station. Furthermore, the dawning of single-stage-to-orbit rocketry may open up the prospect for commercial space pilots to fly between space and low Earth orbit (LEO), taking routes likely to be called the "LEO Run" for short.

However, before you lower your helmet visor to the locked position and get ready to ride a "mountain of fire" into orbit, take note: NASA's semi-yearly astronaut call entails a tough screening process. It is designed to cull the best and brightest. The bottom line: Of the more than 4,000 every-other-year applicants that NASA evaluates, a mere three percent make the first cut. From there, further screening leads to a final 20 selectees, whittled down by an Astronaut Selection Board.

BEING ALL YOU CAN BE

If persistence is what makes you tick —enough to set your sights on an astronaut career — here are the basics. It is worth noting that the majority of astronauts who made the grade had Boy Scout or Girl Scout training in their past. Quite clearly, individuals who were involved in such programs also acquired character-building traits—some of the same qualities NASA is on the look-out for in its astronaut crews.

College preparatory classes in high school are a step in the right direction. Heavy emphasis on math and science-related courses is a must. Striving for the highest grades possible is an obvious condition. Having a clear direction by the the third or fourth year in high school of what specific field you find of greatest interest is critical. Thanks to the variance of fields from which NASA needs astronauts, people skillful in such areas as physical science, engineering, biology and chemistry are also in demand.

While NASA demands a minimum requirement of a bachelor's degree for its astronauts, a majority of those selected have continued their education to the post-graduate level. Beyond a degree, at least three years of professional work experience in a chosen field is mandatory. An advanced degree is desirable and may be substituted for all or part of the experience requirement (i.e, master's degree = one year of work experience, doctoral degree = three years of experience).

KNOCKING ON NASA'S DOOR

NASA selected the first group of U.S. space travelers—the Mercury astronauts—in 1959. From 500 candidates with the required jet aircraft flight experience and engineering training and a height of less than 5 feet 11 inches, seven military men constituted the nation's premier astronaut corps. What does height have to do with ascending to the heavens? As the first American to orbit Earth, astronaut John Glenn, once said: "You didn't fly in the Mercury capsule, you wore it." Space travel has come a long way from single-person, two- and three-seat capsules, to the space plane of today and those of the future.

By 1964, in fact, astronaut requirements had changed, and emphasis was placed on academic qualifications. In 1965, for instance, six scientist astronauts were selected from a group of 400 who had doctorates or equivalent experience in the natural sciences, medicine, or engineering.

The astronaut group named in 1978 was the first of the Space Shuttle flight crews and was composed of 15 pilots and 20 mission specialists. Six of these 35 people were women and four were members of minorities. Since this period of time, seven additional groups have been selected that included a mix of pilots and missions specialists.

Over this stretch of time, hand-picked astronauts have slid into their seats as part of several human spaceflight projects: The Mercury, Gemini, Apollo, the Apollo-Soyuz Test Project (the first joint mission with the then Soviet Union), Skylab, and now the Space Shuttle effort.

NASA accepts applications for the Astronaut Candidate Program from both civilians and military personnel. Current regulations require that preference for appointment to Astronaut Candidate positions be given to U.S. citizens. Three categories of astronauts are in use: pilot astronauts, mission specialists and payload specialists.

Along with the educational background previously outlined, commander/pilot astronaut applicants must also meet the following requirements prior to submitting an application:

- ▲ Have at least 1,000 hours pilot-in-command time in jet aircraft; flight test experience is highly desirable;
- ▲ Must have an ability to pass a NASA Class I space physical, which is similar to a military or civilian Class I flight physical and includes the following specific standards for vision: distance visual acuity—20/150 or better uncorrected, correctable to 20/20 each eye;
- ▲ Height between 64 inches and 76 inches

Pilot astronauts serve as both Space Shuttle commanders and pilots. During the flight, the commander has onboard responsibility for the vehicle, crew, mission success, and safety of flight. The pilot assists the commander in controlling and operating the vehicle and may assist in the deployment and retrieval of satellites using the remote manipulator system—basically, a robot arm.

The mission specialist category, while having similar requirements as the pilot astronaut, also has less stringent guidelines. The qualifying physical is a NASA Class II space physical, which is similar to a military or civilian Class II flight physical. This physical includes the following specific standards for vision: distance visual acuity—20/150 or better uncorrected, correctable to 20/20 each eye. Mission specialist astronauts work with the commander and the pilot and have overall responsibility for coordinating Shuttle operations in the following areas: Shuttle systems, crew activity planning, consumables usage (food and water), and experiment/payload operations. Mission specialists are trained in the details of the shuttle orbiter on-board systems, such as the workings of the robot arm. Additionally, this category of astronaut must be well versed in the all mission objectives and goals, including an understanding of how specific scientific hardware operates. Mission specialists are also assigned space-walking duties—extravehicular activities (EVAs)—perhaps one of the most exhilarating experiences of human space flight.

Lastly, Space Shuttle missions also include payload specialists. The same basic medical standards apply for this category as for a mission specialist. But there is a unique difference in background.

Payload specialists are career scientists or engineers selected by their employer or country for their expertise in conducting a specific experiment or commercial venture. The first payload specialist—basically, the first commercial astronaut—flew in 1984 aboard orbiter Columbia. Charles Walker was a chief test engineer at McDonnell Douglas in St. Louis, Missouri. He went on to fly two more flights in that capacity. His task was to evaluate electrophoresis equipment carried onboard Columbia-hardware that offered the promise of producing ultra-pure and unique "made-in-space" pharmaceuticals.

Payload specialists representing other nations have also ridden into space via the shuttle, including those from Germany, France, Canada, Japan, Mexico, as well as Saudi Arabia.

FLIGHT STATUS

The call from NASA comes. You've passed the tests and you are an astronaut candidate. Now what?

Don't automatically think that the next shuttle mission going up has a seat with your name on it. Some astronauts have waited many years for their first flight. What is in store for you is a tough, demanding training program. Astronaut candidates are trained at the NASA Johnson Space Center in Houston, Texas. Classroom settings provide astronaut candidates with a detailed look at Shuttle systems. Basic science and technology courses involve mathematics, geology, navigation, oceanography and meteorology. The ABCs of rendezvous and docking, astronomy and physics are also among subjects taught.

Part of astronaut training is outside the classroom. Parachute jumping, survival training (be it on water or the land), scuba diving, and working in space suits is another aspect of the workload. Candidates are also exposed to the problems associated with working, living and surviving in varying atmospheric pressures. This is done by training in altitude chambers that mimic space environment conditions.

To simulate microgravity, astronaut candidates board the infamous "Vomit Comet"—a converted KC-135 jet aircraft. Flown on a parabolic trajectory, this airplane can produce periods of microgravity for some 20 seconds.

This airborne roller coaster ride is repeated up to 40 times in a day, creating a weightless environment, like that felt in orbital flight, albeit in short bursts rather than continuous freefall. Why is it called the Vomit Comet? You guess.

Yet another full-body experience with microgravity comes with spending time in the Weightless Environment Training Facility (WETF)—a beefed-up phrase for a large water tank. At the tank's bottom, mockups of the orbiter payload bay and simulated payloads permit astronauts to try out space walking duties. While wearing EVA suits in the WETF, astronauts are made neutrally buoyant, neither rising nor falling in the tank. The result is a close approximation of what a person feels firsthand when working in the space environment.

Part of the year-long instructional program means many hours of time in trainers. Astronaut candidates are tutored in Shuttle mission simulators, associating themselves with all aspects of space flight: prelaunch, ascent, orbit operations, entry, and landing. Several types of simulators are utilized, offering a range of experiences, including procedures for emergency situations. To maintain and sharpen aviator skills, pilot astronauts fly 15 hours per month in NASA's fleet of T-38 jets. Mission specialists fly a minimum of four hours per month.

One very important note: Selection as a candidate does not ensure selection as an astronaut. Final selection is predicated on satisfactory completion of the one-year program. Civilian candidates who successfully complete the training and evaluation are selected as astronauts. They are expected to remain with NASA for at least five years. Successful military candidates are detailed to NASA for a specified tour-of-duty.

HOW IS THE PAY?

Salaries for astronauts start at around $35,000, probably never to rise more than $20,000, regardless of what a person accomplishes. Salaries are based on the Federal Government's General Schedule pay scales for grades GS-11 through GS-14, and are set in accordance with each person's academic achievements and experience

FINAL REENTRY

Space travel after extensive training on the ground is, without question, an incomparable experience. Going from liftoff into orbit takes less than nine thunderous minutes as the rocket flies. High above Earth, a spectacular view awaits all those who free themselves from the planet's one-gravity grip. Upon their return to Earth, astronauts spend several days undergoing medical testing and debriefing. Experiences and lessons learned benefit future crews and the carrying out of upcoming missions.

Public appearances are also a demand on an astronaut's time. Such appearances serve as an important link back to the taxpayer—in a sense, the real mission control for governmental space activities.

Many astronauts have been fortunate enough to be assigned several missions during their NASA employment. Those who leave the astronaut corps typically find high-paying careers in aerospace and non-aerospace professions, or find other governmental niches where they can further contribute their talents.

In considering a career as an astronaut, remember the following—until space travel becomes routine, the number of people who have a career as an astronaut will remain extremely limited. For most, working in space is a dream, not a career. It is best to come up with a backup plan, pursue those interests, and then try to direct your skills toward a mission in space.

YOUR CALL

For additional information regarding a career as an astronaut career visit:
http://www.jsc.nasa.gov/pao/factsheets/

An astronaut application package is yours by writing to:
Astronaut Selection Office
Mail Code AHX
NASA Johnson Space Center
Houston, Texas 77058-3696

APPENDIX A: COMMERCIAL ORGANIZATIONS INVOLVED WITH SPACE

The expansion of telecommunication services (mobile, cellular, fax, data, etc.) has required that the North American Numbering Plan be routinely updated. The following web site lists all past, current, and planned area code changes:

http://www.nanpa.com

For resources listing additional organizations, please refer to the Bookstore at http://www.spacebusiness.com

A.T. Kearney
One Memorial Drive
Cambridge, MA 02142
(617) 374-2600 Fax: (617) 374-4367
http://www.atkearney.com
Management consulting assisting high technology companies

AEC-ABLE Engineering Company
93 Castilian Drive
Goleta, CA 93117
(805) 685-2262 Fax: (805) 685-1369
http://www.aec-able.com
Structures and mechanisms for spacecraft and space platforms

AEM, Inc.
11525 Sorrento Valley Road
San Diego, CA 92121
(619) 481-0210 Fax: (619) 481-1123
http://www.aem-usa.com
Electronics for satellites and other space-related hardware

AeroAstro Corporation
520W Huntmar Park Drive
Herndon, VA 22070
(703) 709-2240 Fax: (703) 709-0790
http://www.aeroastro.com
Small satellites and components, launch vehicles for small payloads

Aerojet
Highway 50
Aerojet Road
Rancho Cordova, CA 95670
(916) 355-1000 Fax: (916) 351-8667
http://www.aerojet.com
Rocket motors and engines, electronic and electro-optic sensors.

Aerospace Corporation, The
2350 East El Segundo Blvd.
El Segundo, CA 90245-4691
(310) 336-5000 Fax: (310) 336-1458
http://www.aero.org
Technical support and consulting in the area of space systems architecture, system studies, specifications definitions, independent analysis and verification. Additional offices in Virginia, Colorado, and New Mexico.

Air Products and Chemicals, Inc.
7201 Hamilton Boulevard
Allentown, PA 18195-1501
(610) 481-4911 Fax: (610) 481-2576
http://www.airproducts.com
Chemical propellant systems and services

Akjuit Aerospace
800 One Lombard Place
Winnipeg, Manitoba R3B 0X3
Canada
(204) 934-5517 Fax: (204) 947-0376
http://www.spaceport.ca
Operates commercial launch facilities at Churchill Rocket Range

Alaska Aerospace Development Corporation
4300 B Street, Suite 101
Anchorage, AK 99503
(907) 561-3338 Fax: (907) 561-3339
Operates commercial launch facilities in Alaska

Allen Osborne Associates
756 Lakefield Road
Westlake Village, CA 91361-2624
(805) 495-8420 Fax: (805) 373-6067
GPS receivers

Alliant TechSystems, Inc.
8400 W 5000 South
Magna, UT 84044
(801) 250-5911
http://www.atk.com
Solid rocket propulsion systems, composite materials

Alliant Techsystems, Inc.
Corporate Headquarters
600 Second St., NE
Hopkins, MN 55343
(612) 931-6000 Fax: (612) 931-6077
http://www.atk.com
Solid rocket propulsion systems, composite materials

AlliedSignal Aerospace
Equipment Systems
1300 West Warner Road
Tempe, AZ 85284
(602) 893-5000 Fax: (602) 893-4922
http://www.alliedsignal.com
Propulsion and power systems for launch vehicles, satellites, and space platforms

AlliedSignal Technical Services Corporation
One Bendix Road
Columbia, MD 21045-1897
(410) 964-7000 Fax: (410) 730-6775
http://www.alliedsignal.com
Ground segment technical support for space missions

AlliedSignal Aerospace Company
AES, Space Environmental Control Systems
2525 West 190th Street
Torrence, CA 90504
(310) 323-9500 Fax: (310) 532-7131
http://www.alliedsignal.com
Systems and subsystems related to guidance and control, and space environmental control systems

AlliedSignal Aerospace Co.
Guidance & Control Systems
699 East Route 46
Teterboro, NJ 07608
(201) 393-2967 Fax: (201) 393-6570
http://www.alliedsignal.com
Spacecraft attitude and control systems, inertial guidance systems for launch vehicles, avionics

Altair Aerospace Corporation
4201 Northview Dr. Suite 300
Bowie, MD 20716
(301) 805-8300 Fax: (301) 805-8122
http://www.altaira.com
Ground/mission command, control, and monitoring software and support

American Mobile Satellite Corporation
10802 Parkridge Blvd.
Reston, VA 20191
(703) 758-6000 Fax: (703) 758-6111
Telecommunications provider

Amptek, Inc.
6 De Angelo Drive
Bedford, MA 01730
(781) 275-2242 Fax: (781) 275-3470
http://www.amptek.com
Electronics, sensors, and instrumentation for spacecraft

Analog Devices
804 Woburn Street
Wilmington, MA 01887
(617) 935-5565 Fax: (617) 937-2000
Electronics

Analytical Graphics, Inc.
325 Technology Drive
Malvern, PA 19355
(610) 578-1000 Fax: (610) 578-1001
http://www.stk.com
Satellite operations software

ANSER
1215 Jefferson Davis Hwy., Suite 800
Arlington, VA 22202
(703) 416-2000 Fax: (703) 416-4451
http://www.anser.org
Non-profit research institute performing technical and business analysis and support

Arrowhead Space & Telecommunications
3040 Williams Drive, Suite 201
Fairfax, VA 22031
(703) 876-4005 Fax: (703) 876-4006
http://www.arrowheadsat.com
Technical telecommunications support and services

Arthur D. Little, Inc.
Technology and Product Development Division
20 Acorn Park
Cambridge, MA 02140-2390
(617) 498-5000 Fax: (617) 498-7045
Technical and management consulting, space hardware design

Astro Aerospace Corporation
6384 Via Real
Carpinteria, CA 93013-2920
(805) 684-6641 Fax: (805) 684-3372
http://www.spar.ca
Manufacture deployable structures, antennas, solar arrays,
and robotic mechanisms

Atlantic Research Corporation
Aerospace Group
5945 Wellington Road
Gainesville, VA 20155
(703) 754-5000 Fax: (703) 754-5316
Manufacture propulsion systems and composite materials structures

Aurora Flight Sciences Corp. of WV
Rt. 1, Box 354
Fairmont, WV 26554
(304) 534-4499 Fax: (304) 534-4498
http://www.hiflight.com
Composite materials

Aydin Corp.
700 Dresher Road
PO Box 349
Horsham, PA 19044
(215) 657-7510 Fax: (215) 657-3830
http://www.aydin.com
Manufacture ground and space equipment for telecommunications

Ball Aerospace & Technologies Corporation
Aerospace Systems Division
1600 Commerce St.
Boulder, CO 80301
(303) 939-4000
Spacecraft optical sensors, cryogenic subsystems, pointing and control hardware, and systems engineering services

Barrios Technology, Inc.
2525 Bay Area Blvd, Suite 300
Houston, TX 77058
(281) 280-1900 Fax: (713) 280-1901
http://www.barrios.com
Engineering and technical services, mission support

BEI Precision Systems and Space Division
1100 Murphy Drive
Maumelle, AR 72113
(501) 851-4000 Fax: (501) 851-5476
Precision angle control components and systems

Betatronix Inc.
110 Nicon Ct
Hauppauge, NY 11788-4289
(516) 582-6740 Fax: (516) 582-6038
Space electronics

BioServe Space Technologies
University of Colorado
Dept. of Aerospace Engineering
Campus Box 429, Boulder, CO 80309
(303) 492-1005 Fax: (303) 492-8883
Biomedical Microgravity experiments and hardware

Boeing Defense & Space Group
Space and Missile Systems Sector
P.O. Box 2515
2201 Seal Beach Blvd.
Seal Beach, CA 90740
(562) 797-3311 Fax: (562) 797-5828
http://www.boeing.com
Seal Beach is the headquarters for Boeing's space activities which has major operating sites in Alabama, Texas, California, and Washington. Major product lines include the manufacture of space systems, space transportation systems, commercial space systems, missile systems, and advanced projects. Boeing is the prime contractor for the Space Shuttle and the International Space Station.

Boeing Space & Missile Systems Sector
Rocketdyne Division
6633 Canoga Avenue
Canoga Park, CA 91303
(818) 586-1000 Fax: (818) 586-2866
http://www.rdyne.rockwell.com
Prime contractor for the Space Shuttle main engines, Space Station electric power systems, liquid rocket booster engines, lasers, optical imaging systems

Boeing Commercial Space Company
P.O. Box 3999
Seattle, WA 98124
(425) 393-1030 Fax: (425) 393-1050
The commercial arm of the Space & Missile Systems Sector

Boeing Space & Missile Systems Sector
499 Boeing Blvd.
PO Box 240002
Huntsville, AL 35824
(205) 461-2805 Fax: (205) 461-2252
Manufacture Space Station hardware, Marshall Space Flight Center and Redstone Arsenal support

Boeing Space & Missile Systems Sector
Space Systems Division
12214 Lakewood Blvd.
Downey, CA 90241
(310) 922-2111 Fax: (310) 922-2032
Design and manufacture Space Shuttle, perform orbiter maintenance, manufacture satellites including Navstar GPS, research and development

Boeing Defense & Space Group
5301 Bolsa Ave.
Huntington, CA 92647
(714) 896-3311
http://www.boeing.com
Manufacture Delta launch vehicle, developing and building major components to Space Station including thermal control, guidance navigation and control, communication and tracking, truss elements.

Booz, Allen & Hamilton, Inc.
7404 Executive Place, Suite 500
Seabrook, MD 20706
(301) 805-5400 Fax: (301) 805-5495
http://www.bah.com
Technical and business services

Brown & Root Services Corp.
4100 Clinton Drive
Houston, TX 77020
(713) 676-5941 Fax: (713) 676-5174
Engineering Services

BRPH Space Facility Architects & Engineers
3275 Suntree Boulevard
Melbourne, FL 32940
(407) 254-7666 Fax: (407) 259-4703
http://www.brph.com
Facilities design and construction

Burns and Roe
800 Kinderkamack Road
Oradell, NJ 07649
(201) 265-2000 Fax: (201) 986-4831
Facilities design and construction

CAL Corporation
1050 Morrison Drive
Ottawa, Ontario K2H 8K7
Canada
(613) 820-8280 Fax: (613) 820-6474
http://www.calcorp.com
Antennas, power modules, electronics, and science instrumentation

CD Radio, Inc.
2175 K Street, NW
Washington, DC 20037
(202) 296-6192
Satellite-based radio broadcasting

Charles Stark Draper Laboratory
555 Technology Square
Cambridge, MA 02139-3539
(617) 258-1000 Fax: (617) 258-3050
http://www.draper.com
Space guidance, navigation, and control systems

Cincinnati Electronics Corp.
7500 Innovation Way
Mason, OH 45040
(513) 573-6100 Fax: (513) 573-6741
http://www.cinele.com
Communications and infrared equipment

Ciprico
2800 Campus Drive
Plymouth, MN 55441
(612) 551-4000 Fax: (612) 551-4002
http://www.ciprico.com
Digital storage devices

Coleman Aerospace Company
5950 Lakehurst Drive
Orlando, FL 32819
(407) 354-0047 Fax: (407) 354-1113
http://www.crc.com
Launch vehicle engines and guidance systems components

Colorado Electronics Corp
3650 North Nevada Ave.
Colorado Springs, CO 80907
(719) 475-0660 Fax: (719) 577-8124
Electronic systems and subsystems for spacecraft

Columbia Communications
7200 Wisconsin Avenue, Suite 701
Bethesda, MD 20814-5228
(301) 907-8800 Fax: (301) 907-2420
http://www.tdrss.com
Telecommunications provider

ComDev Ltd
155 Shelton Drive
Cambridge, Ontario N1R 7H6
Canada
(519) 622-2300 Fax: (519) 622-1691
Electronics

Communications & Power Industries
811 Hansen Way
P.O. Box 51625
Palo Alto, CA 94303
(650) 846-3700 Fax: (650) 424-1744
Electronics for ground facilities

Computer Sciences Corporation
3170 Fairview Park Drive
Falls Church, VA 22042
(703) 876-1000 Fax: (703) 849-1005
http://www.csc.com
Integration, technical support, computer and information services

COMSAT Laboratories
22300 Comsat Drive
Clarksburg, MD 20871
(301) 428-4010 Fax: (301) 428-4600
http://www.comsat.com
Communications technologies and hardware

Comsat Corporation
6560 Rock Spring Drive
Bethesda, MD 20817
(301) 214-3000 Fax: (301) 214-7100
http://www.comsat.com
Communications provider

Creare, Inc.
PO Box 71, Etna Road
Hanover, NH 03755
(603) 643-3800 Fax: (603) 643-4657
http://www.creare.com
Cryogenic equipment

CSP Associates Inc.
55 Cambridge Pkwy.
Cambridge, MA 02142-1234
(617) 225-2828 Fax: (617) 225-2444
Management and market consulting

DATATAPE, Inc.
605 E Huntington Drive
Monrovia, CA 91016-7170
(626) 358-9500 Fax: (626) 358-9100
http://www.datatape.com
Satellite and ground storage devices and data recorders

Datron/Transco Inc.
200 West Los Angeles Ave.
Simi Valley, CA 93065
(805) 584-1717 Fax: (805) 526-3690
http://www.dtsi.com
Satellite telemetry tracking and control systems

DirecTV
2230 E. Imperial Hwy.
El Segundo, CA 90245
(310) 535-5000
http://www.diretv.com
Direct to home television provider

Eagle Picher Industries Inc.
C & Porter Streets
Joplin, MO 64802
(417) 623-8000 Fax: (417) 623-1308
http://www.epi-tech.com
Spacecraft battery and power systems

Earth Resource Mapping
4370 La Jolla Village Drive, Suite 900
San Diego, CA 92122
(619) 558-4709 Fax: (619) 558-2657
http://www.ermapper.com
Remote sensing software and services

Earth Satellite Corporation
6011 Executive Blvd., Suite 400
Rockville, MD 20852-3801
(301) 231-0660 Fax: (301) 231-5020
http://www.earthsat.com
Remote sensing satellite and services

Eaton Corp.
Aerospace Controls Division
2338 Alaska Avenue
El Segundo, CA 90245
(310) 725-3000 Fax: (310) 643-8014
Manufacture spacecraft subsystems hardware

EDO Corporation
Barnes Engineering Division
88 Long Hill Cross Roads
Shelton, CT 06484-0867
(203) 926-1777 Fax: (203) 926-1030
http://www.edocorp.com
Manufacture earth sensors and electronic devices for satellites

EER Systems Corporation
Space Systems Group
10289 Aerospace Rd.
Seabrook, MD 20706
(301) 577-8900 Fax: (301) 794-8884
Software development, systems integration, technical support

EG&G Florida, Inc.
PO Box 21267
Kennedy Space Center, FL 32815
(407) 867-7295 Fax: (407) 867-1199
Facilities operations support and services

EMS Technologies, Inc.
5060 Avalon Ridge Parkway
Technology Park, PO Box 7700
Norcross, GA 30091
(770) 263-9200 Fax: (770) 293-9207
http://www.elmg.com
Microwave systems, antennas, subsystems and components

Engineered Magnetics, Inc.
19300 Susana Road
Rancho Dominguez, CA 90221
(310) 635-9555 Fax: (310) 631-0313
Satellite power systems

ERDAS Inc.
2801 Buford Highway NE, Suite 300
Atlanta, GA 30329-2137
(404) 248-9000 Fax: (404) 248-9400
Remote sensing software and services

ERIM International
PO Box 134008
Ann Arbor, MI 48113-4008
(313) 994-1200 Fax: (313) 994-3890
http://www.erim-intl.com
Institute specializing in remote sensing, robotics and automation, image processing

Essex Corporation
9150 Guilford Road
Columbia, MD 21046
(301) 953-7797 Fax: (301) 953-7880
http://www.essexcorp.com
Simulation and modeling; systems integration; and training systems devices and mockups. Also maintains offices in Huntsville, Alabama.

Frequency Electronics, Inc.
55 Charles Lindbergh Blvd.
Mitchel Field, NY 11553
(516) 794-4500 Fax: (516) 794-4340
Electronic equipment for satellites

Frost & Sullivan
2525 Charleston Road
Mountain View, CA 94043
(415) 961-9000 Fax: (415) 961-5042
http://www.frost.com
Business information research, publishing

Garwood Laboratories
7829 Industry Avenue
Pico Rivera, CA 90660
(562) 949-2727 Fax: (562) 949-8757
http://www.garwoodtestlabs.com
Hardware testing

GE American Communications, Inc.
4 Research Way
Princeton, NJ 08540
(609) 987-4000 Fax: (609) 987-4517
http://www.capital.ge.com
Telecommunications provider

General Sciences Corp.
6100 Chevy Chase Drive, Suite 200
Laurel, MD 20707
(301) 953-2700 Fax: (301) 953-1213
Technical support and integration for ground data and information systems

Geophysical & Environmental Research Corporation
One Bennett Common
Millbrook, NY 12545
(914) 677-6100 Fax: (914) 677-6106
Remote sensing software and services

Globalstar, L.P.
3200 Zanker Road
San Jose, CA 95134
(408) 473-5872 Fax: (408) 473-5548
http://www.globalstar.com
Development and operations of a communications satellite network

GRC International, Inc.
1900 Gallows Road
Vienna, VA 22182
(703) 506-5000 Fax: (703) 903-9431
http://www.grci.com
Engineering and technical analysis and consulting

GTE
1155 S. Telshor Blvd.
Suite 204
Las Cruces, NM 88011-4788
(505) 522-6402 Fax: (505) 521-1957
http://www.gte.com
Support services for ground station operations

Hamilton Standard Space Systems
One Hamilton Road
Windsor Locks, CT 06096-1010
(860) 654-6000 Fax: (860) 654-5515
http://www.hamilton-standard.com
Develop and manufacture environmental control, and life support hardware and systems including space suits, heat exchangers, and pumps

Harris Corporation
Govt. Aerospace Systems Div.
PO Box 94000
Melbourne, FL 32902
(407) 727-5084 Fax: (407) 729-3066
http://www.harris.com
Electronic components including microcomputers and ground terminals

Hernandez Engineering, Inc.
17625 El Camino Real, Suite 200
Houston, TX 77058
(281) 280-5159 Fax: (281) 480-7525
Space flight engineering, support, and technical services

Honeywell, Inc.
Satellite Systems Operations
19019 North 59th Avenue
Glendale, AZ 85308
(602) 561-3000 Fax: (602) 561-3333
http://www.honeywell.com
Data and control mechanisms and control subsystems

Honeywell, Inc.
Space and Strategic Systems Operation
13350 US Highway 19 North
Clearwater, FL 33764
(813) 539-4611 Fax: (813) 539-3447
http://www.honeywell.com
Production of control subsystems and equipment

Hughes Electronics Corporation
P.O. Box 80028
Los Angeles, CA 90080-0028
(310) 568-7200
http://www.hac.com
Corporate Headquarters. Hughes Telecommunications and Space activities include satellite manufacturing (Hughes Space and Communications), satellite ownership and operation (PanAmSat), telecommunications equipment and satellite ground equipment (Hughes Network Systems), direct-to-home television (DirecTV, Inc.)

Hughes Space & Communications Company
P.O. Box 92919
Los Angeles, CA 90009
(310) 364-6000 Fax: (310) 334-4911
http://www.hughespace.com
Design and manufacture communications satellites

Hughes Network Systems, Inc.
11717 Exploration Lane
Germantown, MD 20876
(301) 428-5500 Fax: (301) 428-1868
http://www.hns.com
Satellite ground station networks and equipment

ILC Dover, Inc.
One Moon Walker Road
Frederica, DE 19946-2080
(302) 335-3911 Fax: (302) 335-0762
http://www.ilcdover.com
Pressure suits for hazardous environments

Image Graphics, Inc.
917 Bridgeport Avenue
Shelton, CT 06484
(203) 926-0100 Fax: (203) 926-9705
http://www.igraph.com
Recorders for remote sensing imagery

Instrumentation Technology Assoc., Inc.
15 East Uwchlan Ave., Suite 406
Exton, PA 19341
(610) 363-8343 Fax: (610) 363-8569
Microgravity equipment and services

Integral Systems Inc.
5000 Philadelphia Way, Suite A
Lanham, MD 20706-4417
(301) 731-4233 Fax: (301) 731-9606
http://www.integ.com
Systems integration, software development for satellite command and control

INTELSAT
3400 International Drive, NW
Washington, DC 20008
(202) 944-6800 Fax: (202) 944-7898
http://www.intelsat.int:8080/
Telecommunications provider

International Fuel Cells
195 Governors Highways
South Windsor, CT 06074
(860) 727-2200 Fax: (860) 727-2319
Spacecraft power systems, electrochemical fuel cells

International Launch Services
101 W. Broadway, Suite 2000
San Diego, CA 92101
(619) 645-6400 Fax: (619) 645-6500
http://www.lmco.com/ils
Launch vehicle provider

International Space Brokers, Inc.
1300 Wilson Blvd.
Rosslyn, VA 22209
(703) 841-1334 Fax: (703) 841-0525
http://www.isbworld.com
Satellite and launch insurance provider

International Technology Underwriters
4800 Montgomery Lane, 11th Floor
Bethesda, MD 20814
(301) 654-8585 Fax: (301) 654-7569
Satellite and launch vehicle insurance provider

Interpoint
P.O. Box 97005
Redmond, WA 98073-9705
(425) 882-3100 Fax: (425) 882-1990
http://www.interpoint.com
Power supplies and electronics

Iridium LLC
1575 Eye Street, NW
Suite 500
Washington, DC 20005
(202) 408-3800 Fax: (202) 408-3801
http://www.iridium.com
Telecommunications system developer and provider

Ithaco Space Systems
950 Danby Road, Suite 100
Ithaca, NY 14850
(607) 272-7640 Fax: (607) 272-0804
http://www.newspace.com/ithaco
Satellite control subsystems

ITT Aerospace/Communications Division
1919 West Cook Road, PO Box 3700
Fort Wayne, IN 46801
(219) 487-6000 Fax: (219) 487-6126
http://www.ittind.com
Satellite payloads and sensors

ITT Systems & Sciences
1500 Garden of the Gods Road
Colorado Springs, CO 80907
(719) 599-1500 Fax: (719) 599-1942
Technical support in software and information technology and astrodynamics and mission analysis

Jackson & Tull Aerospace Division
7375 Executive Place, Suite 200
Seabrook, MD 20706
(301) 805-4545 Fax: (301) 805-4538
Engineering support to Goddard Space Flight Center

Johnson Controls
7315 North Atlantic Ave.
Cape Canaveral, FL 32920
(407) 784-7100 Fax: (407) 784-7761
http://www.johnsoncontrols.com
Technical and management support

Kaiser Compositek
1095/1195 Columbia Street
Brea, CA 92821-2928
(714) 990-6300 Fax: (714) 990-2681
http://www.kaisercompositek.com
Manufacture composite subsystems

Kearfott Guidance and Navigation Corp.
150 Totowa Road
Wayne, NJ 07474-0946
(973) 785-6000 Fax: (973) 785-6025
http://www.kearfott.com
Guidance and navigation systems and electronics

Kistler Aerospace Corporation
3760 Carillon Point
Kirkland, WA 98033
(425) 889-2001 Fax: (425) 803-3303
http://www.kistler.com
Reusable launch vehicles

Kollmorgen Motion Technologies Corp.
501 First Street
Radford, VA 24141
(540) 639-9045 Fax: (540) 731-4193
http://www.kollmorgen.com
Motion control electronics and systems

KPMG Peat Marwick
Commercial Space & High Technology Group
2001 M Street, NW
Washington, DC 20036
(202) 467-3250 Fax: (202) 293-5437
http://www.kmpg.com
Management, marketing and policy services

KRUG Life Sciences Inc.
1290 Hercules Drive, Suite 120
Houston, TX 77058
(281) 212-1200 Fax: (281) 212-1211
Biomedical services

L-3 Communications
Space & Satellite Control
1150 Academy Park Loop, Suite 240
Colorado Springs, CO 80910
(719) 637-8200 Fax: (719) 637-8501
Software products and services for defense space activities

L-3 Telemetry and Instrumentation
15378 Avenue of Science
San Diego, CA 92128
(619) 674-5100 Fax: (619) 674-5145
http://www.ti.lmco.com
Ground hardware and software

Launchspace
7929 Westpark Drive, Suite 100
McLean, VA 22102
(703) 749-2324 Fax: (703) 749-3177
http://www.launchspace.com
Publishing, training

Litton Industries, Inc. Amecom Division
5115 Calvert Road
College Park, MD 20740
(301) 864-5600 Fax: (301) 864-4934
Electronic hardware and subsystems

Litton Systems, Inc.
Guidance and Control Systems
5500 Canoga Avenue
Woodland Hills, CA
(818) 715-4040 Fax: (818) 715-2019
http://www.littongcs.com
Guidance and control systems for spacecraft and launch vehicles

Litton PRC
1500 PRC Drive
McLean, VA 22102
(703) 556-1000 Fax: (703) 556-1174
http://www.prc.com
Technical, engineering, and software services and support

Lockheed Martin Corporation
Corporate Headquarters
6801 Rockledge Drive
Bethesda, MD 20817
(301) 897-6000 Fax: (301) 897-6083
http://www.lmco.com
Launch vehicle and satellite manufacturer, information systems, technical & operations support. Major facilities in Florida, Colorado, California, Texas, Maryland

Lockheed Martin Skunk Works
1011 Lockheed Way
Palmdale, CA 93599
(805) 572-2974 Fax: (805) 5772-4163
http://www.lmco.com
Design and manufacture of X-33 single-stage-to-orbit launch vehicle

Lockheed Martin Management and Data Systems
935 First Ave.
King of Prussia, PA 19406
(610) 531-7400 Fax: (610) 531-0006
http://www.lmco.com
Command and control systems, gateway earth stations, voice compression

Lockheed Martin Missiles & Space
1111 Lockheed Way
Sunnyvale, CA 94089
(408) 742-7151 Fax: (408) 742-5798
http://www.lmms.lmco.com
Manufacture commercial, military, and civil satellites and components

Lockheed Martin Astronautics
P.O. Box 179
Denver, CO 80201-0179
(303) 977-3000 Fax: (303) 971-4902
Manufacture space launch systems, space systems, and ground systems

ManTech International Corp.
12015 Lee Jackson Hwy.
Fairfax, VA 22033-3300
(703) 218-6000 Fax: (703) 218-6068
Engineering and technical support services
Major operations in Colorado, Virginia, California, and Maryland

Mobile Communications Holdings, Inc.
1133 21st NW, Suite 550
Washington, DC 20036
(202) 466-4488 Fax: (202) 466-4493
http://www.ellipso.com
Developing a satellite system for telecommunications

MOOG Space Products Division
P.O. Box 18
East Aurora, NY 14052-0018
(716) 652-2000 Fax: (716) 687-4466
http://www.moog.com
Satellite components

Motorola, Inc.
Satellite Communications Division
2501 South Price Road
Chandler, AZ 85248
(602) 441-3033 Fax: (602) 441-7028
http://www.mot.com
Manufacture mobile communication and wireless receivers, manufacture and integrate satellites and satellite constellations such as Iridium and Celestri

MPB Corporation
Precision Park, P.O.Box 547
Keene, NH 03431
(603) 352-0310 Fax: (603) 355-4553
http://www.timken.com
Bearings for space systems

Nichols Research Corp.
4040 S Memorial Parkway
Huntsville, AL 35815-1502
(205) 883-1140 Fax: (205) 880-0367
http://www.nichols.com
Engineering and technical services and support
Major facilities in Virginia, Alabama, California

OAO Corp.
7500 Greenway Center Drive
Greenbelt, MD 20770
(301) 345-0750 Fax: (301) 345-0952
http://www.oao.com
Engineering and technical services and support

Odetics Space Division
1585 South Manchester Ave.
Anaheim, CA 92802
(714) 758-0100 Fax: (714) 780-7649
http://www.odetics.com
Data Recorders and Electronics

Omitron, Inc.
6411 Ivy Lane, Suite 600
Greenbelt, MD 20770
(301) 474-1700 Fax: (301) 345-4594
http://www.omitron.com
Engineering and technical services and support

Orbcomm Corporation
21700 Atlantic Blvd.
Dulles, VA 20166
(703) 406-6000 Fax: (703) 406-3504
http://www.orbital.com
Telecommunications provider

Orbital Sciences Corporation
21700 Atlantic Blvd.
Dulles, VA 20166
(703) 406-5000 Fax: (703) 406-3509
http://www.orbital.com
Satellite and launch vehicle manufacturer for internal and external communications and remote sensing activities. Major facilities in Virginia, Maryland, California, and Arizona

Orbital Technologies Corporation
Space Center, 1212 Fourier Drive
Madison, WI 53717
(608) 827-5000 Fax: (608) 827-5050
Technical services and research

PanAmSat
One Pickwick Plaza
Greenwich, CT 06830
(203) 622-6664 Fax: (203) 622-9163
http://www.panamsat.com
Telecommunications provider

PCI Remote Sensing Corp.
1925 N Lynn Street, Suite 803
Arlington, VA 22209
(703) 243-3700 Fax: (703) 243-3705
http://www.pci.on.ca
Remote sensing services

Phillips Business Information
1201 Seven Locks Road, Suite 300
Potomac, MD 20854
(301) 340-1520 (301) 309-9473
http://www.phillips.com
Publishing, media

Physical Sciences Laboratory
New Mexico State University
PO Box 30002
Las Cruces, NM 88003-0002
(505) 522-9100 Fax: (505) 522-9434
http://www.psl.nmsu.edu
Operations and testing facilities support

Pratt & Whitney
Chemical Systems Division
PO Box 49028
San Jose, CA 95161-9028
(408) 776-9121 Fax: (408) 779-4202
Develop and manufacture propulsion systems

Pratt and Whitney
Government Engines and Space Propulsion
PO Box 109600
West Palm Beach, FL 33410-9600
(561) 796-2000 Fax: (561) 796-7258
Develop and manufacture propulsion systems

Pressure Systems Inc.
6033 E Bandini Blvd.
City of Commerce, CA 90040
(213) 722-0222 Fax: (213) 721-6002
Manufacturing hardware for launch vehicles and satellites

Qualcomm Inc.
6455 Lusk Blvd.
San Diego, CA 92121
(619) 587-1121 Fax: (619) 658-2501
http://www.qualcomm.com
Manufacture communications hardware and mobile/cellular telephones

Raytheon Company
141 Spring Street
Lexington, MA 02173
(781) 862-6600 Fax: (781) 860-2520
http://www.raytheon.com
Electronics and communications equipment

Raytheon E-Systems
1501 72nd Street North
St. Petersburg, FL 33710
(813) 381-2000 Fax: (813) 302-4418
Electronics

Raytheon Engineers & Constructors
5555 Greenwood Plaza Blvd., Suite 100
Englewood, CO 80111
(303) 843-2600 Fax: (303) 843-2169
Facility design and construction

Research Systems Inc.
2995 Wilderness Place, Suite 203
Boulder, CO 80301
(303) 786-9900 Fax: (303) 786-9909
Data visualization software

Schaeffer Magnetics
A division of Moog, Inc.
9175 Deering Avenue
Chatsworth, CA 91311-5897
(818) 341-5156 Fax: (818) 341-3884
Spacecraft subsystems manufacturer

Scientific Atlanta
4311 Communications Drive
Norcross, GA 30093
(770) 903-5000 Fax: (770) 903-3902
http://www.sciatl.com
Manufacture telecommunication and electronic equipment

Sea Launch Company
Southcenter South
PO Box 3999, MS 8X-50
Seattle, WA 98124
(425) 393-1030 Fax: (425) 393-1050
http://www.boeing.com/sealaunch
Launch vehicle provider

Servo Corporation of America
123 Frost Street
Westbury, NY 11801
(516) 938-9700 Fax: (516) 938-9644
http://www.servo.com
Small satellite components

Southwest Research Institute
Instrumentation & Space Research Division
6220 Culebra Road
San Antonio, TX 78238
(210) 522-2748 Fax: (210) 647-4325
http://www.swri.com
Electronics and components for satellites

Space Electronics Inc.
4031 Sorrento Valley Blvd.
San Diego, CA 92121
(619) 452-4167 Fax: (619) 452-5499
http://www.newspace.com/spaceelec
Space electronics

Space Hardware Optimization Technology
5605 Featherengill Road
Floyd Knobs, IN 47119
(812) 923-9591 Fax: (812) 923-9598
http://www.shot.com
Microgravity research, hardware, and support

Space Imaging EOSAT
1207 Grant Street, Suite 500
Thornton, CO 80241
(303) 254-2000 Fax: (303) 254-2215
http://www.spaceimage.com
Remote sensing data provider and services

Space Systems/Loral
3825 Fabian Way
Palo Alto, CA 94303
(650) 852-4000 Fax: (650) 852-7912
Satellite manufacturer

SPACEHAB, Inc.
1595 Springhill Road
Suite 360
Vienna, VA 22182
(703) 821-3000 Fax: (703) 821-3070
http://www.spacehab.com
Microgravity research facility provider

Spaceport Systems International L.P.
3769-C Constellation Road
Lompoc, CA 93436
(805) 733-7370 Fax: (805) 733-7372
http://www.calspace.com
Manages the commercial spaceport near Vandenberg AFB

SPAR Aerospace Ltd
21025 Trans Canada Highway
Ste Anne de Bellenue, Quebec H9X 3R2
Canada
(514) 457-2150 Fax: (514) 457-2724
http://www.spar.ca
Satellite manufacture, robotic systems, antennas, electronics

Spectrum Astro Inc.
1440 North Fiesta Blvd.
Gilbert, AZ 85233
(602) 892-8200 Fax: (602) 892-2949
http://www.spectrumastro.com
Satellite manufacturer, space hardware including electrical power subsystem

Steven Myers & Associates, Inc.
4695 MacArthur Court
Newport Beach, CA 92660
(714) 975-1550 Fax: (714) 975-1624
http://www.smawins.com
Proposal and technical writing support

Storm Control Systems
2214 Rock Hill Road
CIT Tower, Suite 502
Herndon, VA 20170-4214
(703) 478-6200 Fax: (703) 478-6670
Software

Sverdrup Technology Inc.
620 Discovery Drive
Huntsville, AL 35806
(205) 971-0100 Fax: (205) 971-9475
Engineering, technical, and facilities support and services

Tecstar, Inc.
15251 Don Julian Road
City of Industry, CA 91745-1002
(818) 968-6581 Fax: (818) 336-8694
http://www.appliedsolar.com
Satellite power systems. Additional facilities in North Carolina.

Teledesic Corporation
2300 Carillon Point
Kirkland, WA 98033
(425) 602-0000 Fax: (425) 803-1404
http://www.teledesic.com
Satellite communications provider

Teledyne Brown Engineering
Cummings Research Park
PO Box 070007
Huntsville, AL 35807
(205) 726-5724 Fax: (205) 726-6286
http://www.tbe.com/services/space/space.html
Technical support and services, microgravity

Thiokol Corporation
2475 Washington Blvd.
Ogden, UT 84401-2398
(801) 629-2270 Fax: (801) 629-2251
http://www.thiokol.com
Propulsion systems

Trimble Navigation, Ltd
645 North Mary Avenue
Sunnyvale, CA 94086
(408) 481-8000 Fax: (408) 481-2997
Global Positioning System hardware

TRW Space and Electronics Group
One Space Park
Redondo Beach, CA 90278
(310) 812-4321 Fax: (310) 812-7110
http://www.trw.com
Satellite manufacturing, payload design, avionics, power and propulsion, optics, electronic devices, testing, etc.

TSI/Telsys
7100 Columbia Gateway Drive
Suite 150
Columbia, MD 21046
(410) 872-3900 Fax: (410) 872-3901
http://www.tsi-telsys.com
Ground station communications systems

United Space Alliance
1150 Gemini
Houston, TX 77058
(281) 212-6000 Fax: (281) 212-6179
http://www.unitedspacealliance.com
Manages the ground operations and integration for the Space Shuttle

U.S. Aviation Underwriters, Inc. (USAIG)
One Seaport Plaza
199 Water Street
New York, NY 10038
(212) 952-0100 Fax: (212) 344-1381
Satellite and launch vehicle insurance

Vacco Industries
10350 Vacco Street
South El Monte, CA 91733
(626) 442-6943 Fax: (626) 442-6943
Propulsion system hardware

Vision International
5301 Shawnee Road
Alexandria, VA 22312
(703) 658-4000 Fax: (703) 658-4021
Remote sensing and image processing. Divisions in Colorado focus on remote sensing and launch vehicle software.

Wyle Laboratories, Inc.
7800 Highway 20 West
PO Box 077777
Huntsville, AL 35807-7777
(205) 837-4411 Fax: (205) 721-0144
http://www.wylelabs.com
Technical support, services, and testing

APPENDIX B: KEY SPACE TERMS AND ACRONYMS

- Antenna—A device for transmitting and receiving radio waves. Depending on its use and operating frequency, antennas can be a single piece of wire, a grid of wires, a sophisticated parabolic shaped dish or an array of electronic elements
- Apogee—The point at which a spacecraft is at its farthest point from the earth's surface
- Astrobiology—Study of life on planets
- Astronautics—The science of space travel, including the building and operating of space vehicles
- Attitude—The position of a spacecraft as determined by the inclination of its axis to some point of reference
- Attitude control—The orientation of the satellite in relationship to the earth and sun
- Bandwidth—A measure of spectrum frequency use or capacity.
- Bioastronautics—Study of the effects of space travel on plant or animal life
- Bus—The electronic brain of a satellite
- Celestial mechanics—Study of orbital paths of celestial bodies under the influence of gravitational fields
- COTS Software—Commercial off-the-shelf software
- Cryogenic—Generally refers to liquids or the use of liquids at super cold temperatures.
- DBS—Direct Broadcasting Services, such as direct-to-home television or radio

- Delta V—Velocity changes that enable a space vehicle to change its trajectory
- DoD—Acronym for the U.S. Department of Defense
- Downlink Data—The frequency range utilized by the satellite to retransmit signals down to earth for reception
- DTH—Acronym for Direct-to-Home. Refers to direct broadcast satellite television and radio services
- Earth Station—Term used to describe the combination of an antenna, low noise amplifier, down-converter, and receiver electronics used to receive a signal transmitted by a satellite.
- Eccentric orbit—An elliptical orbit, one with a very high apogee and a low perigee
- Eclipse—When a satellite passes through the line between the earth and sun or earth and moon
- Equatorial Orbit—An orbit with a plane parallel to the earth's equator
- Escape Velocity—The velocity a vehicle must attain in order to overcome the gravitational field.
- ESA—Acronym for European Space Agency
- Expendable Launch Vehicle (ELV)—A rocket that can only be launched once and whose components cannot be refurbished for future use
- Fairing—The area of the launch vehicle where a payload is attached until its release into orbit.
- FCC—Acronym for the Federal Communications Commission
- FEMA—Acronym for the Federal Emergency Management Administration
- Footprint—A map of the signal strength showing the power contours of equal signal strengths as they cover the earth's surface.
- Frequency—Number of times that an alternating current goes through its complete cycle in one second of time.
- Frequency Coordination—A process to eliminate frequency interference between different satellite systems or between terrestrial microwave systems and satellites
- FSS—Fixed Satellite Services—Refers to services using satellites in geostationary orbit
- Fuel cell—Miniature electric power plants
- GEO—Acronym for Geostationary Orbit

- Geostationary Orbit—The point in space at which an object will revolve at the same speed as a point on earth. From the earth, the object appears to be stationary.
- GIS—Acronym for Geographical Information Systems
- GPS—Acronym for Global Positioning Satellite system
- Guidance, Navigation, and Control (GN&C)—System that measures the velocity and directory of the spacecraft, compares it with its memory, and issues commands for corrections
- Hypersonic—Speeds faster than Mach 5 (five times the speed of sound)
- Inclination—The angle of an orbit. Equatorial orbit has an inclination of zero, a polar earth orbit has an inclination of 90 degrees.
- Insurance—Multiple types of coverage exist to reduce financial risks related to the satellite, launch, and on-orbit operation
- Integration—Technical activity regarding the combining of different systems and/or components, such as placing a satellite inside the launch vehicle
- Ka-band—Frequency range from 18 -31 GHz
- Ku-band—Frequency range from 10.9—17 GHz
- L-band—Frequency range from 0.5—1.5 GHz
- Launch Vehicle—A vehicle that is used to deliver payloads from the Earth into space. Also see Expendable Launch Vehicle and Reusable Launch Vehicle
- LEO—Acronym for Low Earth Orbit
- Man-rated/Human-rated—Equipment which is considered reliable enough to be used by people.
- Mobile Satellite Services (MSS)—Satellite-based mobile telecommunications systems that provide coverage over extended areas including in remote locations.
- Mux/Demux—A multiplexer combines several different signals (e.g. video, audio, data) onto a single communication channel for transmission. Demultiplexing separates each signal at the receiving end.
- NASDA—Acronym referring to the organization leading Japan's national space program
- NOAA—Acronym for the National Oceanic and Atmospheric Administration
- Orbit—The path of a body under the influence of gravitational or other force around another body.

- Orbital Period—The time it takes to complete one orbit
- Orbital Velocity—The speed required to establish and maintain a spacecraft in orbit.
- Payload—The cargo that is being carried for a mission
- Perigee—The point in an orbit at which the spacecraft is closest to the earth.
- Period—The amount of time that it takes a satellite to complete one revolution of its orbit
- Polar Orbit—An orbit with its plane aligned parallel with the polar axis of the Earth
- R&D—Acronym for Research and Development
- RDDT&E—Acronym for Research, Design, Development, Test, and Evaluation
- Receiver—An electronic device that enables a particular satellite signal to be separated from all others being received by an earth station, and converts the signal into a format for video, voice, or data
- Remote Sensing—The collection of data from a distant location. Remotely-sensed data is typically collected from sensors located on an aircraft, balloon, or spacecraft
- Reusable Launch Vehicle (RLV)—A rocket that after placing its payload in space, is returned to the Earth and refurbished for future flights.
- RF—Acronym for radiofrequency
- Rocket—See launch vehicle
- Satellite—A sophisticated electronic communications relay station.
- Sensor—An electronic device that monitors or collects data
- Slot—The longitudinal position in geostationary orbit at which a communications satellite is located
- Sounding Rocket—A research rocket which sends equipment into the upper atmosphere, takes measurements, and returns to the ground
- Specific Impulse—A means of expressing rocket performance
- Spectrum—Range of electromagnetic radio frequencies used in the transmission of voice, data, and video
- Spin-off—Commercial or other benefits derived from space research
- Support Services—Includes technical and business support activities, such as legal and licensing, finance, consulting, and publishing

- Telemetry—The technique whereby information on the health, activity, and location of a spacecraft is measured and transmitted to the ground
- Teleport—A facility that provides uplink and downlink services
- Three Axis Stabilization—Type of spacecraft stabilization in which the body maintains a fixed attitude relative to the orbital track and the earth's surface.
- Transfer Orbit—A highly elliptical orbit that is used as an intermediate stage for placing satellites in geostationary orbit
- Transponder—A combination receiver, frequency converter, and transmitter package that is a physical part of a communications satellite
- Turnkey—A system that is supplied, installed, and sometimes managed by one vendor or manufacturer
- Uplink Data—Earth to space communications pathway
- USAF—Acronym for the United States Air Force
- Value Added Provider—An organization that modifies an existing product or data source before delivering it to the end customer. In remote sensing, value added providers manipulate the raw data into more usable forms, such as a graphical map.
- VSAT—Acronym for very small aperture terminal. Refers to small earth stations. Many are used to connect remote locations with a central home office computer, such as those for ATM bank machines, credit card processing, etc.
- Window—Limited time period during which a space vehicle can be launched if it is to accomplish its mission

INDEX

Comments, Suggestions Improvements

Have you learned something in your job search that you would like to share with others?

Have you come across information that you believe we should include in future editions?

Do you have changes to material presented in this book?

Perhaps you would like to send a note letting us know how you used this book.

We would like to hear from you.

Please send your letters to:

Space Publications
P.O. Box 5752
Bethesda, MD 20824-5752
United States

sacknoff@spacebusiness.com
http://www.spacebusiness.com